ELDERBERRY

OUR MEDICINE CHEST

Is Elderberry Syrup an Effective Anti-Flu Remedy? Does it Work? What Does the Science Say?

Now with introductions for both consumers and medical professionals

WJ FARRELL

Table of Contents

List of Figures

List of Tables

Preface

Were the glowing customer testimonials of our product the result of a placebo effect or did elderberry really possess healing power? In the summer of 2017, I still had so many questions.

Since 2009 we've successfully grown elderberry on our farm in Lyme, Connecticut and in 2014 my spouse Liz and I launched an elderberry product that contains only elderberry and apples. Thankfully, our customers rave about our product for its taste, simplicity, lack of preservatives, etc. but in truth, I know many of them take elderberry because they believe in or expect health benefits.

By law, we can't promote the health benefits of elderberry and, in fact because of the law we withhold publishing on our website hundreds of customer reviews that suggest elderberry has helped them.

But what's the real story? Are the blogs and internet "news" stories true? Does elderberry work? Do the results of scientific studies conducted over the last thirty years support the views of many of my customers? Is there scientific consensus about some or all of elderberry's medical benefits? I started this research project so that I could learn what proof science offered about elderberry's health benefits and in order to make good farming / business / health decisions for my young family.

I didn't have the scientific education to evaluate the studies directly (plus some studies cost $300 each) so I began connecting with scientifically trained, credentialed people who could conduct literature reviews and answer my questions. My original plan was to seek answers from a few people on a limited set of questions. But within six months or so I developed an insatiable appetite for more answers. Over the course of a year-and-one-half, I hired eight researchers at considerable expense.

Please understand that I directed the researchers to write using easy to understand, non-technical language so that this information could be available to people of all education levels. Although the articles were edited by me for typos and the like, they are very much the work of each contributor. It was important to me that I give each author the autonomy to develop their own conclusions based on their independent review of the scientific studies. I did not hire a professional editor to mediate between you and the researchers answering my questions. I hope the benefit of raw makes up for any lack of polish.

Structure of the Book

The introduction is divided into two parts: one for the average consumer, and another for the medical professional. Written by a Doctor of Veterinary Medicine, these provide an excellent overview of how elderberry works in the body and the degree to which its health claims are scientifically substantiated. If you can give this book only 20 minutes, you should read one or both of these sections.

The <u>first section</u> of the book describes the basic components of the berry (and the flower) that help our bodies. Written plainly, with lucid explanations of the bigger words like "anthocyanins" and "polyphenols," this section forms the foundation of the rest of the book.

The <u>second section</u> jumps into the "how" of the berry – how to extract the most nutrition from it, how to evaluate the HYPE around elderberry by looking at its edible parts, and a quick primer on why the whole food is better than the sum of the parts of the food.

The <u>third section</u> examines elderberry and the flu, colds, diseases, and health conditions. Perhaps the most technical part of Elderberry: Our Medicine Chest, these articles summarize and evaluate the medical research about how elderberries help our bodies. Again, our goal was to stay away from any HYPE and ask researchers to stick to a careful evaluation of the science.

Lastly, in <u>section four</u> we 1) bring you important answers about more specific topics, such as possible side-effects and potential interactions with medications or health conditions, 2) review the evidence to answer the big "how much should I take" question (dosage), and also 3) summarize current evidence about whether elderberry supplements or foods are safe for children. You might not find satisfactory answers to all your questions, but please keep in mind that I asked researchers to use science rather than speculation to answer my questions.

<u>Maybe you have a pet and would like to provide them with elderberry supplementation?</u> At the beginning and end of this book we provide links to download our free e-book "Elderberry and Pets – What does the Science Say" and ask you to become an email subscriber for free updates on the latest elderberry science. Make sure to visit our website www.elderberryguru.com to learn about the history and lore surrounding this ancient medicinal plant and for updates as scientists advance our understanding of the elder plant.

Sincerely,

WJ Farrell

Elderberries Benefit People - But What About Pets?

What's Inside?

Over 14 pages of clearly written, well-researched conclusions about elderberries's documented impacts on animal health.

Topics covered include:

- *Has Elderberry Been Studied In Animals?*
- *Does is Work? In What Ways?*
- *What Dose is Acceptable?*
- *Are side-effects possible?*
- And more!

Introduction for Consumers

By Jon Stack, DVM

This book is intended to serve as a comprehensive guide to the elderberry plant and to provide information and insight about the history, nutritional qualities, and potential health applications of the elderberry plant.

Each article is intended to provide detailed information about certain aspects of elderberries, exploring many of the proposed qualities of elderberry plants and products from a scientific standpoint and identifying what questions about elderberries merit further exploration. This FAQ seeks to answer some of the most basic questions about elderberries, and should serve to provide a launching point for the more detailed articles on the website.

What is Elderberry?

From the most basic standpoint, Elderberry "trees" are actually a group of deciduous, flowering shrubs which are classified under the scientific genus Sambucus. There are multiple species found around the world, and most species produce fruits which are edible when properly prepared. The most well-known and cultivated species, the European black or blue Elder, is classified as Sambucus nigra.

Although European in origin, the black Elder grows widely across the midwestern and eastern United States and is commonly called the American Black Elder. In the western parts of North America, Canada, and Mexico, several other species of Elderberry shrubs thrive.

Elderberry trees have a long history of use and many cultures have regarded the plants as having medicinal qualities. The fruit and flowers have been made into wines, pastries, teas, and cosmetics and used as a panacea for maladies ranging from fevers, skin infections, nausea, and dry skin. The pigments of the berries have also been used as dye for clothing and hair.

Is Elderberry a Food, Supplement, a Drug, or All Three?

If one simply performs an internet search looking for information about elderberry, the volume of information returned can be overwhelming. One will find recipes for elderberry pies, jams, and wines. You will also likely find recommendations to either consume elderberry regularly, either in the form of juices, foods or as manufactured supplements.

Look even further and you will find there are recommendations for its use against infections and other maladies. The question then becomes what is the best use of elderberry, and why does it seem to have such a broad range of applications?

Elderberries and elder flowers are often used in foods, and are perfectly safe when consumed after proper preparation. They are both a rich source of anti-oxidants (specifically, polyphenols). It is worthwhile to note that elderberry fruit and elder flowers differ in the types of anti-oxidants they contain, which may make it worthwhile to consume both components.

Many people consume elderberries simply because they enjoy the taste of the fruit or the products. The polyphenolic content of elderberries makes them an appealing supplement for those who are trying to enjoy the proposed health benefits of diets rich in polyphenols, which is associated with reduced risks of cancer, cardiovascular disease, inflammation, and degenerative disorders.

Going even further, there is evidence that elderberries have medicinal properties that are on par with or better than the best available anti-influenza virus medications. Certain compounds found in elderberries appear to directly inhibit the ability of the influenza virus to gain entry into host cells, and other compounds may actually activate the immune system against the virus.

A variety of supplements have been created specifically for use in upper respiratory infections, and in Europe, elderberry products are regulated in a way which acknowledges their traditional use as a medicinal compound. Rather than trying to classify elderberry as a food, supplement, or drug, it is best to consider it a versatile berry which likely has uses for all three situations.

Are All Elderberry Products Created Equal?

Now that we are discussing the potential use of elderberries as a supplement or remedy for influenza, it is important to note that there are numerous elderberry products which are marketed for their purported anti-viral properties or immune-boosting properties. Many of these compounds contain other ingredients besides elderberries, which complicates the questions even further, and there is no direct head-to-head comparison to determine if any one product out-performs the others.

A few commercially available products have been used in peer-reviewed trials against the flu with promising results, but most have no published information regarding their efficacy. At this time, there is also no information regarding what is an appropriate dose of elderberry, either as a supplement or as an anti-viral remedy.

There are some major factors to consider when selecting an elderberry product. These factors include the actual concentration of elderberry in the product (as some products may contain very little actual elderberry) and what is the form of the elderberry (such as an extract, juice, or jam).

The more processing which has gone into the elderberry product, the more likely it is that the beneficial ingredients will be degraded. All products are not created equal, and there are varying levels of clinical evidence for the use of certain products in situations like respiratory infections. The chapter "Which Elderberry Products are Best for the Flu?" provides more detailed information on specific products along with quality rankings.

What are the Health Benefits of Taking Elderberry?

A large part of elderberry's historical fame is due to the wide range of purported health benefits. It is important to keep in mind that there are varying degrees of evidence for each of the health claims, but many are supported by some initial scientific research.

The most robust claims with the highest levels of support are for the use of elderberries against influenza, upper respiratory, and sinus infections. Multiple studies have shown reductions in the severity or the duration of symptoms against both viral and bacterial infections.

Elderberry may also have more general health benefits, especially against inflammatory conditions. Inflammation is a hallmark of a diverse range of disease processes, including obesity, arthritis, and many chronic gastrointestinal diseases.

Elderberry has been shown to be helpful in reducing markers of inflammation in the body associated with these diseases. Over time, chronic inflammation can lead to increasing dysfunction, immune dysregulation, and loss of function of the affected body systems. Reducing inflammation may help improve long-term health in these scenarios.

There are many other studies evaluating the benefits of elderberries. It has been shown to be an effective laxative with few side effects, and may have application in cancer, cardiovascular disease, and in the treatment of wounds. These benefits are all aside from the more general benefits which are believed to be associated with diets rich in polyphenols.

Can I take Elderberry Every Day?

In general, elderberry is safe to consume on a daily basis. There are caveats to this, of course. When taken as a supplement, one needs to consider all the ingredients it may contain besides elderberry. There also needs to be some level of consideration if you are currently taking medications for other health conditions, as it is possible that it may alter the efficacy or absorption of some medications.

The amount of elderberry should also be considered. Over-consumption may lead to high daily calorie intake, particularly if the method of ingestion is through products like pastries or juices/jams containing added sugars. Within moderation, however, daily consumption of elderberries or supplements is generally considered safe and a good option for obtaining polyphenols and other anti-oxidants in which Western diets are often deficient.

Is Taking Elderberry More Effective than Getting a Flu Shot?

It is important to distinguish the difference between elderberry and a flu shot. A flu shot is intended to induce protective immunity by exposing the immune system to viral compounds, allowing it to produce anti-bodies ahead of an infection, which then offers the potential to completely prevent or severely mitigate an infection.

The efficacy of the flu shot varies from year to year due to the wide number of influenza viruses and the need for scientists and vaccine manufacturers to make an "educated guess" about which strains are most likely to spread around the world.

Some years, the vaccines may be extremely protective, while in others, it may only confer partial or very limited protection.

Elderberries and elderberry supplements have mostly been evaluated in people who have already been infected with influenza. It is not considered a preventative, but there is evidence that it can shorten the duration and severity of influenza infections. Elderberry also seems to have immune-boosting properties.

Rather than skipping a flu shot and relying on elderberry, it is best to think of them as complementary strategies towards treating the flu. An effective flu shot is the best preventative, but those who develop the flu, or in the years when the flu shot is less efficacious, using elderberry after infection may help you recover more quickly and help boost the immune system to fight off the infection.

How Much Elderberry Should I Take All at Once and for How Long?

The ideal dose of elderberry for supplemental or medicinal purposes is not completely clear, but the most recent clinical trials have used between 300 to 500 mg of elderberry compound per day for influenza infections.

There are other studies which have used other dosing ranges and product formulations, however, and some of these used either certain volumes of juice per day or a certain weight of berries from which juice was extracted.

There is no strong data to support any of the dosing schemes which have been used. When using supplements, it is best to follow the dosing guidelines listed on the product inserts or containers, as it is likely the safe dose which has been evaluated by the manufacturer.

If you are taking elderberry supplements for an illness, it is reasonable to continue taking the supplements until you have recovered from your symptoms. At least one study used a juice preparation for 5 weeks with no significant adverse effects.

Does Elderberry Have Health Benefits besides Fighting the Flu?

Although elderberry is achieving some fame for its use in the setting of influenza, there is evidence that it has other benefits as well. It appears to have some use in infections besides influenza, particularly the common cold and bacterial sinus infections. For both of these situations, elderberry appears to reduce the severity and duration of symptoms.

Elderberry also seems to have more broad uses as well; it has shown great promise as an anti-inflammatory agent in obesity, arthritis, Crohn's disease, and gingivitis. It is also undergoing extensive research for potential benefits in regulating cholesterol, as an anti-cancer agent, and for promoting wound-healing.

Can I Give Elderberry to My Kids? If So, How Much?

Currently, there is no information regarding the safety of elderberry when given to children for medicinal purposes. Although elderberries are safe when properly prepared as food items, many supplements intended for fighting infections or for other health concerns contain a variety of ingredients, and not all of these may be safe for children.

All the available studies show that it is safe for adults who are not currently pregnant or breast-feeding. If you intend to give elderberries to your children, the safest approach is as a food item, and not a supplement, particularly if it contains multiple ingredients.

Can I Give Elderberries to My Pets?

Although there have been some studies about the use of elderberries (or compounds found in elderberries) in animals, most of these studies have been in lab animals such as rats and mice. There are no studies evaluating the use of elderberries in the more popular pets like dogs and cats.

Holistic veterinarians do sometimes prescribe elderberry products for pets with signs of upper respiratory infections, and at least one veterinary toxicologist believes it to be generally safe, but it is still recommended to consult with a veterinarian before using it in your pets. Some products may contain other ingredients (such as artificial sweeteners, garlic, onions, and grapes) which can be surprisingly toxic to dogs and cats.

What's the Most Potent Form of Elderberry?

If you are looking for elderberry products which contain the highest concentrations of anti-oxidants and polyphenols, juices and jams tend to contain the highest concentration of these compounds. When it comes to efficacy against infections such as the flu however, there are a few commercial products which have high concentrations of elderberries and have scientific evidence supporting their use against infections.

Sambucol is the most well-researched product, while Rubini also has supporting data, but is not currently available in the United States. There are other products which contain varying levels of elderberry, but all of these products also contain additional ingredients which may contribute to their efficacy. There is some evidence that jams and jellies may be just as effective as commercial products.

Can I Get Sick if I Take Too Much Elderberry?

The major risks from ingestion of elderberries stems from consumption of unripe berries, barks, stems, or leaves from the plant. Each of these parts of the plant contains compounds which can generate cyanide, a serious toxin. Out of this concern, it is important to cook the ripe berries for at least 30 minutes at 150 degrees before consumption. This inactivates the compounds and renders the berries safe to eat.

The plants do contain other potentially toxic compounds called ribosome inactivating proteins (RIPs). These are a natural defense mechanism which protects the plant from insects, and there is evidence that these compounds may be harmful to isolated human cells grown in labs, but overall there is not a great deal of evidence showing harm to humans, but they do have the potential to cause allergic reactions in some people.

As elderberries seem to have impacts on the immune system, immune-compromised individuals may want to exercise more caution, as elderberry may have more unpredictable effects in these patients.

In general, however, most people are likely to tolerate elderberries when consumed as a normal part of a diet. Avoid taking higher-than-recommended doses of supplements, as the other ingredients in these products could potentially be harmful. Always ensure that foods prepared with elderberries have been thoroughly cooked.

Can Elderberry Conflict with My Other Medications?

Although there are no specific studies evaluating the combination of elderberries and prescription medications in people, there are theoretically possible interactions.

People who are on anti-diabetic medications should exercise caution, as elderberries may reduce blood sugar levels on their own, and this may increase the risk for hypoglycemia (low blood sugar). Elderberry may also reduce the effectiveness of some opioid pain medications or potentiate the affects of some anesthetic drugs.

People receiving immunosuppressive or immuno-modulatory drugs may also want to avoid elderberry products, as there may be unpredictable impacts on the immune system. Finally, those who are taking elderberry as a remedy for constipation may want to avoid combining it with other laxatives.

Is it Safe for Me to Eat Elderberries Raw?

Consumption of raw elderberries and elderberry products should be avoided, as most of the components of elderberry shrubs contain compounds which are converted into cyanide in the body. There are no reported human deaths from eating raw elderberry products, but people have become quite sick and have required hospitalization.

Livestock which has consumed large amounts of the shrubs have died from cyanide poisoning, however.

Cooking these compounds will inactivate the toxic substances, but higher concentrations are found in the bark, leaves, and stems, and these portions may require prolonged exposure to heat. The exception to this is the flowers, which do not contain toxic substances. These may be used in teas and pastries without concern for specific preparation. Avoid incorporating any stems into foods, however, as the stems do contain high levels of toxic compounds.

Does Elderflower Have as Many Health Benefits as Elderberry?

Although the elderberries receive most of the attention in regards to health benefits, elderflowers are also a good source of polyphenols – and the ones found in elderflowers may be even more potent anti-inflammatories compared to the ones found in the berries.

Furthermore, these polyphenols appear to be stable when made into a tea, which is one of the most common ways elderflowers are enjoyed. By consuming both elderflowers and elderberries, you can enjoy a broad spectrum of polyphenols which complement each other very well.

Summary

These are just some of the more common questions people may have regarding elderberries and elderberry products. More detailed information answering each of these questions will be provided in the upcoming chapters. As you will see, there is a wealth of information regarding elderberries and there are many, many more questions which need further research.

Introduction for Medical Professionals

By Jon Stack, DVM

There is a growing body of scientific evidence that elderberries and elderberry products have medicinal benefits. The majority of evidence supports their use in the settings of influenza, upper respiratory, and sinus infections; however, there is growing evidence that elderberries have other beneficial properties as well.

Many physicians and nurses may find that their patients seek out natural remedies to maladies, and that patients are likely to come across elderberry-based products during their search. This chapter is intended to serve as a guide for physicians and other medical professionals faced with patients who have questions about elderberry products.

How safe are elderberries for my patients?

The question of safety is dependent on the product, the preparation, the intended use, and of course, the individual patient. In general, prepared elderberry products are safe for most people. It is important to keep in mind that the majority of the elderberry plant is, in fact, toxic if consumed raw. Raw leaves, stems, bark, unripe berries, and roots all contain cyanogenic glycosides as well as alkaloids and ribosome inactivating proteins (RIPs).

The cyanogenic glycosides are similar to those found in bitter almonds and can be metabolized to hydrocyanic acid. In the body, this metabolism is typically slow and rarely causes life-threatening signs in humans, although a person can become ill with gastrointestinal signs. Cyanogenic glycosides are heat labile and the toxic parts of the elderberry plant can be rendered safe by cooking.

The RIPs are of uncertain significance. They do not appear to result in clinical illness in humans, although cell cultures demonstrate some toxicity to mammalian cells.

Their most significant effect appears to be as a potential allergen.

The most commonly consumed parts of the elderberry plant are the berries and the flowers. The flowers and ripe berries themselves contain no (or low) toxins and are safe to consume raw. It is generally recommended to cook the berries and remove any stems, leaves, or other parts of the plant for at home preparation. When properly cooked, elderberries can be made into wines, juices, pastries, jams, and other foods and are completely safe for consumption.

A number of commercial products geared towards supplementation or herbal remedies now exist. The majority of these products are made from extracts and processing the berries makes them safe. In general, elderberries appear to be safe, particularly if supplemented in the diet as foods. Commercial products may contain a variety of ingredients which may also need to be considered in the context of the individual patient.

Although elderberries appear safe, special considerations for pregnant and nursing women as well as young children should always be made. There are not currently any safety studies addressing the safety of elderberries or commercial elderberry products in these groups. Although elderberry-based foods are likely safe, the commercial products may carry uncertain risks.

What are the indications for using elderberry?

From a clinical standpoint, the use of elderberries most supported by clinical evidence is for influenza, upper respiratory, or sinus infections. Clinical trials using various elderberry products have rather consistently demonstrated a shortened duration and severity of illness using such products.

Elderberries have also demonstrated clinical efficacy in reducing gingivitis and as a laxative during constipation. There are many other uses in which elderberries are also believed to be helpful, including as a general anti-inflammatory agent, in the management of hyperlipidemia, and potentially in wound-healing. The research in these other areas is quite preliminary, and some is based on animal models rather than humans.

Outside of specific health concerns, elderberries and elderflowers seem to contain a mixture of polyphenols, particularly anthocyanins, which are potent free-radical scavenging compounds. A major indication for elderberries may simply be as a dietary addition to increase the intake of polyphenols, which are normally part of a diet rich in fruits and vegetables.

How does elderberry work against influenza infections?

Elderberries appear to work against influenza through several different mechanisms of action. The fruit naturally contains lectins which bind to mammalian cell surface sialic acid receptors. These receptors are also utilized by influenza viruses in order to gain entry into the cell. In this mechanism, it appears that the lectins simply out-compete the viral particles for these receptors, blocking entry.

A variety of other pathogens also depend on sialic acid residues as part of the infectious process, including Streptococcus pneumoniae, several Haemophilus species, and E. coli. Elderberries have not yet been evaluated against these pathogens, but there may be a theoretical benefit for these diseases as well.

Other mechanisms which have been demonstrated in cell-culture studies show that at least two flavonoids in elderberries are also capable of directly binding to the virus particles themselves. This appears to render them incapable of entry into mammalian cells. At least 10 strains of influenza have been shown to be antagonized by elderberry extracts.

Elderberries also appear to have impacts on the immune system by affecting cytokine secretion. An in vitro study has found that human monocytes treated with elderberry extracts produce higher levels of IL-1beta, TNF-alpha, IL-6, and IL-8 compared to stimulation with the highly immunogenic lipopolysaccharide (endotoxin).

A study in mice found that treatment with elderberry extract during influenza infection also resulted in increases in circulating antibodies and secreted IgA antibodies. The exact mechanisms by which elderberry extracts exert these effects is unclear, but they may play a role in immune stimulation during infections.

How do elderberry products compare to antiviral medications such as Tamiflu, Relenza, or Rapivab?

The only direct head-to-head comparison which has been performed is an elderberry and echinacea based extract compared against Tamiflu in patients infected with influenza. The patients who received the elderberry/echinacea product showed a faster recovery with fewer side effects compared to Tamiflu, which may cause nausea and vomiting.

Elderberry flavonoids, like Tamiflu, bind to influenza virus particles. Unlike Tamiflu, however, elderberries may also have immune-stimulating properties as well as by binding to sialic acid receptors on cell surfaces. This may confer a multi-modal advantage over Tamiflu, which has encountered resistance in some strains of influenza. Both appear to shorten the duration of illness compared to placebo.

No direct comparison has been made between elderberry products and Relenza or Rapivab. The major side effect of Relenza is bronchospasm and may reduce the course of illness by an average of 1.5 days (compared to the 2-4 days with elderberries and Tamiflu). Rapivab has been found to be an effective drug in some cases of influenza, but it tends to be reserved for those patients who do not respond to Tamiflu or Relenza.

Compared to any of these medications, elderberry products are generally inexpensive. For patients who do not require hospitalization, elderberry-based products may be a reasonable option if they would not normally require a specific anti-viral medication.

For those patients who become critically ill and require hospitalization, use of anti-viral medications should be prioritized until more research is available on the efficacy of elderberries.

How strong is the clinical evidence for elderberry products against influenza?

There have been five clinical trials using elderberry-based products in the context of influenza infections. The trials themselves have been relatively small (containing between 25 - 312 patients). Three of the trials evaluated a product called Sambucol, which contains elderberry extract. Although all of the Sambucol trials were small, they were all randomized, double-blind, placebo-controlled trials which yielded similar, statistically significant results. All three trials reflected shorter recovery times compared to the placebo-treated groups and an absence of reported side effects.

A fourth trial evaluated a different elderberry product. This trial was also a randomized, double-blind, placebo-controlled trial and found similar results of faster recovery times, lessening of symptoms, and no reported side effects.

The final trial (n = 312) evaluated yet a different product, but found that symptom severity and illness duration were significantly reduced and shortened, respectively. This trial also attempted to evaluate a protective effect of elderberries against infection, but found the infection rates were the same in both the treatment and control groups.

While the studies have been generally well-designed and consistent, the biggest critique is that they have been small. Larger clinical trials may provide more rigorous evaluation of elderberries and elderberry products; however, these small trials have been promising. All trials reflect apparent safety of the products in the treatment groups.

Are there known interactions with any medications?

Unfortunately, there are no human studies at this time regarding interactions between elderberries and other medications. Some of elderberries' known properties may raise concern for some medications, and patients should be appropriately counseled on possible interactions if they have concurrent health conditions.

Elderberries seem to have a regulatory effect on blood glucose, and it is possible that they could have an insulin-sparing effect. There is a theoretical possibility that diabetics could experience episodes of hypoglycemia if adding elderberries to their health regimen, although this has not been shown.

Several animal studies have been performed which may raise concern for the interaction of elderberries with other medications as well. Elderberries may interfere with analgesic action, reducing the efficacy of some opioid medications. It also appeared to cause resistance to phenobarbital as an anesthetic induction agent. The mechanisms behind these effects are unclear, as is whether they translate to humans, but elderberries may be advised against for patients who take phenobarbital for issues such as seizures.

Elderberries' apparent effects on the immune system may also be potentially concerning for those who are either dealing with autoimmune diseases, organ transplants, or are immunosuppressed for other reasons. If immune-activation is a property of elderberries, it is possible that it carries some risk of adverse reactions in these patients.

What are the possible side effects of elderberry use?

The side effects of elderberries appear to be generally minimal, with most clinical trials assessing elderberry products reporting no side effects. One exception is the use of elderberry tea as a laxative. In that study, elderberry tea did appear to increase the risk of abdominal cramping in constipated individuals (similar to the effects of most over-the-counter laxatives). Due to their high fiber content, over-indulgence in elderberry fruit may also result in diarrhea or gastrointestinal upset.

Other specific potential side-effects are likely to depend on what additional ingredients the supplements may contain. In general, side effects with elderberry supplements are not expected, but, as described under the "side effect" question, there may be certain scenarios in which elderberries could have unexpected effects.

Are there standardized dosing regimens for elderberry?

Currently, there is no known "dose" for elderberries. Commercial products typically have dosing guidelines which are presumably based on their own internal product development process. The most recent human trials have generally used between 300-500 mg of elderberry extract per day. The products used in studies tends to vary widely, however. The content and qualities of extracts and ingredients also likely varies significantly from study to study.

This is an area in which advising patients can become difficult. It is best to discuss individual products with your patient, and come to a consensus together about whether or not the product fits into a treatment plan. Most products are likely to be safe when consumers follow the labeled dosage, however the bigger question is in regards to the efficacy.

If patients are trying to maximize the amount of polyphenols in their diet, the best approach may be simply to add in elderberries into the diet. Jams and juices tend to contain the highest concentration of these compounds, and are safe to consume.

Can elderberries replace any standard treatments?

Right now, the quality and quantity of evidence does not give elderberry products equal standing with drugs and techniques which have undergone FDA-approved clinical trials. Elderberries may have certain uses in uncomplicated respiratory or sinus infections and are likely an excellent source of polyphenols, however standards of care should always be adhered to.

Patients may like to incorporate elderberries into their treatment regimens, which should be considered on a case-by-case basis. For most patients, this should not be a major issue.

Elderberries can be an excellent adjunctive therapy or a simple home remedy for minor maladies which are typically self-limiting in nature. Until more research becomes available, elderberries should be regarded as a generally safe, generally beneficial source of polyphenols. These polyphenols appear to have many qualities which support overall wellness, and perhaps someday there will be more research about their other effects in the body.

Section 1: What's in an Elderberry? Antioxidants, Polyphenols, and More!

The Key to Elderberry's Health Benefits

By Melissa M. Melough, RD

Research has consistently confirmed the health benefits of a diet rich in fruits and vegetables.(1,2) However, researchers are still working to address many unanswered questions regarding the exact components of fruits and vegetables that promote health as well as the precise ways in which they act in the body. One prevailing theory suggests that the health-promoting properties of fruits and vegetables may be attributable in part to their antioxidant content.

Antioxidants are substances that protect against harmful free radicals. They are rich in fruits and vegetables where they play a key role in protecting the plant from pathogens and harsh conditions such as UV radiation.(3)

It is believed that consuming antioxidants can confer the same type of benefits to humans, protecting us from damage caused by free radicals produced through normal metabolic activities and as a consequence of environmental pollutants or other stressors. Because of these potential benefits, researchers have put considerable time and effort into determining which foods rank highest in antioxidant content.

Photo Credit: Shutterstock

Numerous laboratory tests exist to measure the antioxidant power of foods including the ORAC (oxygen radical absorbance capacity) and TEAC (Trolox equivalent antioxidant capacity) tests. However, all of these tests share one primary flaw: they are all conducted in test tubes and therefore cannot shed much light on how a food's antioxidants might behave in the human body.

Furthermore, the free radicals used in these tests are different from the ones that occur naturally in the human body.(4) Antioxidant tests are also highly influenced by the experimental conditions such as the chemicals used, the temperature, and the reaction timing.(5) These tests are rarely standardized, so the results collected by one researcher may be different from those of another researcher.

Additionally, results of these tests are often expressed in units that are difficult to interpret and compare. Therefore, many researchers agree that antioxidant "scores" are often misleading and are not very useful for ranking the healthfulness of food.

Ready for some good news? Although there is a lot left to learn about antioxidants, studies conducted in both humans and animals have demonstrated that consuming antioxidant-rich foods helps to reduce the damage caused by free radicals.(6–9)

Importantly, numerous large studies have indicated that taking supplements of the antioxidants vitamin C and E was not effective for preventing cardiovascular disease or cancer.(10–12) This suggests that eating foods rich in antioxidants and other nutrients is more beneficial for health than consuming isolated antioxidants taken in a capsule. Furthermore, the health benefits provided by antioxidant-rich foods are probably determined by other factors besides their total antioxidant capacity.

Phenolic compounds are one important class of plant-based nutrients that are found in antioxidant-rich foods. This class of nutrients includes flavonoids, anthocyanins, flavones, isoflavones and several other groups. Phenolic compounds act as antioxidants, but they also play other important roles in the body. Studies have indicated that certain phenolic compounds may reduce inflammation, prevent the formation of blood clots that can lead to heart attack or stroke, and improve blood pressure by relaxing blood vessels.(13–15)

Phenolic compounds are found in almost all edible plants including grains, fruits, vegetables, herbs, and spices. Berries contain a wide range of phenolic compounds often in relatively high concentrations compared to other foods.(16) Elderberries stand out as one example of a berry rich in phenolic compounds.

One study that examined wines made from different fruits showed that elderberry wine was higher in total phenolic content than wines made from blueberry, black currant, raspberry, cranberry, and several others.(17)

In fact, of all the wines tested, elderberry wine was second only to a traditional wine made from red grapes. Other researchers compared the contents of several individual phenolic compounds in 18 different kinds of fresh berries.(16) Their results showed that elderberry ranked highly in several compounds including quercetin (a flavonoid) and cyanidin (an anthocyanidin).

Table 1: Phenolic Content Comparison Among Different Berries

FRUIT	PHENOLIC CONTENT*			
	CAFFEIC ACID	QUERCETIN	CYANIDIN	P-COUMARIC ACID
Elderberry	179	331	3316	18
Chokeberry	832	348	8421	60
Cranberry	65	207	379	11
Blueberry	325	95	214	11
Bilberry	111	81	2933	65
Black Currant	27	50	1452	39

*Content of various phenolic compounds (mg/kg of fresh weight) adapted from data presented by Määttä-Riihinen et al. in the Journal of Agricultural and Food Chemistry (2004)(16).

Thousands of different phenolic compounds have been identified in plant foods, and much more research is needed to determine which ones are most important for human health. Until more is known, ranking foods in terms of their phenolic content likely gives an imperfect picture of the overall healthfulness of different foods.

While it is clear that phenolic compounds have powerful health-promoting effects and are essential components of the diet(18), we'll have to wait for conclusive evidence about the what, when, and why of antioxidant health benefits.

REFERENCES

1. Tuso, P. J., Ismail, M. H., Ha, B. P. & Bartolotto, C. Nutritional Update for Physicians: Plant-Based Diets. Perm. J. 17, 61–66 (2013).

2. Dinu, M., Abbate, R., Gensini, G. F., Casini, A. & Sofi, F. Vegetarian, vegan diets and multiple health outcomes: A systematic review with meta-analysis of observational studies. Crit. Rev. Food Sci. Nutr. 57, 3640–3649 (2017).

3. Pandey, K. B. & Rizvi, S. I. Plant polyphenols as dietary antioxidants in human health and disease. Oxid. Med. Cell. Longev. 2, 270–8 (2009).

4. Schaich, K. M., Tian, X. & Xie, J. Hurdles and pitfalls in measuring antioxidant efficacy: A critical evaluation of ABTS, DPPH, and ORAC assays. J. Funct. Foods 18, 782–796 (2015).

5. Shahidi, F. & Zhong, Y. Measurement of antioxidant activity. J. Funct. Foods 18, 757–781 (2015).

6. Noratto, G. D., Chew, B. P. & Atienza, L. M. Red raspberry (Rubus idaeus L.) intake decreases oxidative stress in obese diabetic (db/db) mice. Food Chem. 227, 305–314 (2017).

7. Lara-Guzmán, O. J., Álvarez-Quintero, R., Osorio, E., Naranjo-Cano, M. & Muñoz-Durango, K. GC/MS method to quantify bioavailable phenolic compounds and antioxidant capacity determination of plasma after acute coffee consumption in human volunteers. Food Res. Int. 89, 219–226 (2016).

8. Kolarzyk, E., Pietrzycka, A., Zając, J. & Morawiecka-Baranek, J. Relationship between dietary antioxidant index (DAI) and antioxidants level in plasma of Kraków inhabitants. Adv. Clin. Exp. Med. 26, 393–399 (2017).

9. Abu Hafsa, S. H. & Ibrahim, S. A. Effect of dietary polyphenol-rich grape seed on growth performance, antioxidant capacity and ileal microflora in broiler chicks. J. Anim. Physiol. Anim. Nutr. (Berl). 1–8 (2017). doi:10.1111/jpn.12688

10. Sesso, H. D. et al. Vitamins E and C in the Prevention. J. Am. Med 301, 2123–2133 (2010).

11. Klein, E. A. et al. Vitamin E and the Risk of Prostate Cancer. Jama 306, 1549 (2011).

12. Cook, N. R. et al. A randomized factorial trial of vitamins C and E and beta carotene in the secondary prevention of cardiovascular events in women: results

from the Women's Antioxidant Cardiovascular Study. Arch. Intern. Med. 167, 1610–1618 (2008).

13. Xia, E. et al. Biological Activities of Polyphenols from Grapes. Polyphenols Hum. Heal. Dis. 1, 47–58 (2013).

14. Morton, L. W., Abu-Amsha Caccetta, R., Puddey, I. B. & Croft, K. D. Chemistry and biological effects of dietary phenolic compounds: relevance to cardiovascular disease. Clin. Exp. Pharmacol. Physiol. 27, 152–159 (2000).

15. Kong, J. M., Chia, L. S., Goh, N. K., Chia, T. F. & Brouillard, R. Analysis and biological activities of anthocyanins. Phytochemistry 64, 923–933 (2003).

16. Määttä-Riihinen, K. R., Kamal-Eldin, A., Mattila, P. H., González-Paramás, A. M. & Törrönen, R. Distribution and contents of phenolic compounds in eighteen scandinavian berry species. J. Agric. Food Chem. 52, 4477–4486 (2004).

17. Rupasinghe, H. P. V. & Clegg, S. Total antioxidant capacity, total phenolic content, mineral elements, and histamine concentrations in wines of different fruit sources. J. Food Compos. Anal. 20, 133–137 (2007).

18. Balasundram, N., Sundram, K. & Samman, S. Phenolic compounds in plants and agri-industrial by-products: Antioxidant activity, occurrence, and potential uses. Food Chem. 99, 191–203 (2006).

A Closer Look at Elderberry's Polyphenols

By Susan Elrod, PhD

Many consumers are interested in the content of polyphenols, the wonderful antioxidants suggested to improve health, in their favorite foods. The good news if you love elderberries and other dark fruits is that, in general, the stronger the color of the food, the more polyphenols (or other phytochemicals).

This makes sense: it is these very compounds that give these foods their color. The beautiful deep color of elderberries is due to a set of compounds known as anthocyanins (1); these compounds make fruits and vegetables red, blue, or purple, depending on their structure.

Elderberries have been studied for various potential health benefits, including anti-inflammatory effects (2), and it is hypothesized that such potential benefit may be due to the high amount of polyphenols in elderberry, particularly anthocyanins.

That stated, studies have also shown the possibility of significant variation in the phenolic and anthocyanin content according to the type of elderberries and where and when they were grown (1).

This means that it may not be possible to predict the amount of polyphenols you're consuming in your fruits. However, you can still be confident that brightly-colored fruits and vegetables will likely have significant amounts of polyphenols, and fruits like elderberries will primarily contain anthocyanins. So what are anthocyanins, anyway? And what can they do for you?

Anthocyanins are a specific type of polyphenol; you'll often hear them grouped with anthocyanidins. The reason for this is that anthocyanins are just anthocyanidins with a sugar group added. So then what are anthocyanidins? They have the following basic structure:

Figure 1: Chemical Structure of Anthocyanidins

The name of the anthocyanin will be based on what molecules are attached to each point labeled "R." For our purposes, you don't need to understand the differences between these molecules or compounds; the important thing here is that a lot of those R-groups will have oxygen, which reacts with free radicals in our body to protect against a lot of damage.

The specific anthocyanins/anthocyanidins in elderberries include delphinidin, cyanindin, peonidin, and pelargonidin (1). These compounds have been studied for potential health benefits, particularly for prevention of certain cancers (3), cardiovascular disease (4), and osteoarthritis (5).

Nothing has been proven, but these compounds found in elderberries and other fruits and vegetables, to varying extents, are especially beneficial in health and the prevention of certain chronic diseases. But even with all the evidence regarding the potential health benefits of anthocyanins, these compounds have not necessarily been proven to be better for you than other polyphenols.

Most polyphenolic phytochemicals have been studied for the prevention of chronic disease and other health benefits. A lot of that is due to the inherent antioxidant properties of polyphenols, but there have been other mechanisms suggested for polyphenolic benefits besides antioxidant effects.

That's why it's generally recommended that we not only increase our fruit and vegetable intake but that we consume a variety of fruits and vegetables. According to the World Health Organization, "… fruits and vegetables are rich sources of vitamins and minerals, dietary fiber and a host of beneficial non-nutrient substances including plant sterols, flavonoids, and other antioxidants and consuming a variety of fruits and vegetables helps ensure an adequate intake of many of these essential nutrients." (6)

So while you shouldn't put all your polyphenolic eggs in the anthocyanin basket (so to speak), if you are interested in increasing your anthocyanin intake, those dark purple elderberries are an excellent source.

How about the polyphenol content of processed snacks made with fruits and vegetables? There has been some limited research done on those fruit and vegetable chips you'll find in the health food section of your grocery store.

Again, these studies are limited, but it appears chips made from red or purple potatoes (which, in their raw form, will also be high in anthocyanins! (7)) do have reasonable polyphenolic and anthocyanin content, though these compounds decrease dramatically in the final product due to the cooking and frying process (8).

Photo Credit: Shutterstock

So while those colorful veggie chips might be a better alternative to your usual potato chips, a lot of the polyphenols may have disappeared during preparation. Plus you have to balance that polyphenolic content against increased fat and salt in such prepared foods. Another study looking at fruit and vegetable chips found that snacks made from chokeberries and blackcurrants, which are dark in color like elderberries, had higher phenolic content and free radical scavenging ability than chips made from apples or carrots (9).

Still, both of these studies reported dramatically lower phenolic and anthocyanin values than that measured for raw elderberries(10).

REFERENCES

1. Wu, H., et al. Determination of anthocyanins and total polyphenols in a variety of elderberry juices by UPLC-MC/MC and other methods. Acta Hortic 2015; 1061:43-51.

2. Farrell, N., et al. Anthocyanin-rich black elderberry extract improves markers of HDL function and reduces aortic cholesterol in hyperlipidemic mice. Food Funct 2015; 6:1278-1287. Farrell, N., et al. Black elderberry extract attenuates inflammationand metabolic dysfunction in diet-induced obese mice. Br J Nutr 2015; 114:1123-1131.

3. Thomasset, S,. et al. Do anthocyanins, cancer chemopreventive pigments in the diet, merit development as potential drugs? Cancer Chemother Pharmacol 2009; 64:201- 211.

4. Dayoub, O., et al. Pleiotropic beneficial effects of epigallocatechin gallate, quercetin, and delphinidin on cardiovascular diseases associated with endothelial dysfunction. Cardiovasc Hematol Agents Med Chem 2013; 11:249-264.

5. Haseeb, A., et al. Delphinidin inhibits IL-1β-induced activation of NF-κB by modulating the phosphorylation of IRAK-1(Ser376) in human articular chondrocytes. Rheumatology (Oxford) 2013; 52:998-1008.

6. http://www.who.int/elena/titles/fruit_vegetables_ncds/en/

7. Harborne, J. B., Plant polyphenols. 1. Anthocyanin production in the cultivated potato. Biochem J 1960; 74:262-269.

8. Nemś, A., et al. Anthocyanin and antioxidant activity of snacks with coloured potato. Food Chem 2015; 175:175-182.

9. Gramza-Michałowska, A. & Człapka-Matyasik, M. Evaluation of the antiradical potential of fruit and vegetable snacks. Acta Sci Pol Technol Aliment 2011; 10:63-72.

10. Duymus, H. G., et al. In vitro antioxidant properties and anthocyanin compositions of elderberry extracts. Food Chem 2014; 115:112-119.

The Elderberry - "The Tree of Music" - is a Rich Source of Anti-oxidants

By Sarah Crawford, PhD

The American elderberry (Sambucus canadensis L.) derives its name from the Anglo-Saxon "aeld" which means fire, as its hollow stems were used to blow air into firepits to increase their flame. The Genus name "Sambucus" means wind instrument because the branches were used by Native Americans to make flutes, from which it earned the name, "the tree of music."

The elderberry is a perennial shrub native to the eastern part of North America with delicate white blooms that mature to deep purple berries in late summer to fall. The first reports of its cultivation (or planting and harvesting) in America date back to 1761, although the first cultivars (those developed for agricultural use), called Adams 1 and Adams 2, were only presented in 1926. The elderberry fruit is used to prepare fragrant and delicious jams, juice, wines, and pies.

This rich summer fruit was also used by Native American cultures as a medicinal, a use that may be appropriate as more recent research has demonstrated that the elderberry is rich in phytochemicals with anti-oxidant properties, including phenolics and anthocyanins. Although not as well-known as the more famous blueberry and raspberry, analyses of its phytochemical makeup show that the anti-oxidant properties of the elderberry compare extremely well to these small fruit relatives.

Anti-oxidant Content of Elderberries

Phenolic compounds are important because they are scavengers of free radicals-highly reactive molecules that damage DNA and cells of the body.

Free radical scavengers - also called anti-oxidants - target and neutralize free radicals to prevent them from damaging the cell. It is well known that plant phenolics are excellent anti-oxidants, so researchers measure the amounts of phenolics with free radical scavenging ability in the cell.

Scientists have determined that both the phenolic (see Table 2) and anti-oxidant capacity (see Table 3) of elderberries is comparable to that of blackberries and other small berry fruits. (1)

In fact, the darker the pigmentation of the fruit, the greater its anti-oxidant capacity, because the pigmentation reflects the amount of anthocyanins – a group of free radical scavenging phenolics- in the plant. Moreover, the anti-oxidant capacity of berries is significantly higher than almost any other food group (see Table 4). (2) not as well-known as the more famous blueberry and raspberry, analyses of its phytochemical makeup show that the anti-oxidant properties of the elderberry compare extremely well to these small fruit relatives.

Photo Credit: Shutterstock

The National Food and Nutrient Analysis program of the US Department of Agriculture ranked berries #1 in anti-oxidant content per serving among 1,113 foods tested. (2) For this reason, plant developers are attempting to produce elderberry cultivars with even higher levels of these important anti-oxidants.

In fact, clinical research studies have shown that anthocyanins show preventive effects for degenerative diseases associated with aging- not surprising since much research has shown that the aging process is associated with accumulated effects of free radical damage to cells of the body. (3)

Research by Thole et al. (4) has shown that elderberries of both American (S. canadensis) and European (S. nigra) species are potent inhibitors of cyclo-oxygenase-2, an enzyme that triggers inflammatory processes linked to many types of cancer, including colon cancer. These species of elderberry were also found to activate quinone reductase, an important biomarker for cancer prevention. (4)

Conclusion

Plants with dark, small berries are among the richest of all foods in phenolic content and associated anti-oxidant properties. Elderberries are one of the most important members of this group, based on research showing that the levels of these important phytochemicals are among the highest of any berry.

Important clinical research suggests that the anti-oxidant effects of these phenolic compounds may play a critical role in the prevention of cardiovascular disease, cancer, and degenerative disorders. (3,4)

Table 2: Phenolic Content of Berry Fruits (1)

FRUIT	PHENOLIC CONTENT [MG GALLIC ACID·100 G -1 F.W.]
Blackberry	247.25±11.14
Elderberry	535.98±8.04
Blueberry	424.72±14.05
Wild Strawberry	165.46±13.07

Table 3: Anti-oxidant Levels of Berry Fruits (1)

FRUIT	FRAP [MM FE·100G-1 F.W.] ANTIOXIDANT MEASUREMENT
Blackberry	15.17±0.21
Blueberry	30.48±1.42
Elderberry	29.56±1.92
Wild Strawberry	10.99±0.29

Table 4: The Comparative Anti-oxidant Content of Foods(2)

PRODUCT CATEGORY	ANTIOXIDANT CONTENT (MMOL/100 G)
Spices and herbs	0.803–125.549
Nuts and seeds	0.029–13.126
Chocolates and sweets	0.092–10.474
Vegetables and vegetable products	0.018–4.694
Ready-to-eat cereals	0.157–4.291
Desserts and cakes	0.000–4.097
Berries and berry products	0.978–4.059
Fruit and fruit juices	0.081–2.512

Table 4: The Comparative Anti-oxidant Content of Foods (Cont.)(2)

PRODUCT CATEGORY	ANTIOXIDANT CONTENT (MMOL/100 G)
Beverage	0.000–2.135
Soups, sauces, gravies, dressings, etc	0.000–1.566
Fast food	0.001–1.262
Infant foods and beverages	0.017–1.248
Legumes	0.008–1.184
Snacks	0.148–1.170
Grains and grain products	0.009–0.997
Dairy products	0.011–0.763
Mixed-food entrees	0.026–0.731
Fats and oils	0.187–0.531
Meat, meat products, and substitutes	0.052–0.509
Poultry and poultry products	0.072–0.388
Fish and seafood	0.025–0.141
Egg and egg dishes	0.009–0.047

REFERENCES

1. Jabłońska-Ryś E, Zalewska-Korona M, Kalbarczyk J, Anti-oxidant capacity, ascorbic acid and phenolics content in wild edible fruits. Journal of Fruit and Ornamental Plant Research Vol. 17/(2) 2009: 115-120.

2. Halvorsen BL, Carlsen MH, Phillips KM, et al. Content of redox-active compounds (i.e., antioxidants) in foods consumed in the United States. Am J Clin Nutr. 2006 Jul;84(1):95-135.

3. Youdim KA, Martin A, Joseph JA. Incorporation of the elderberry anthocyanins by endothelial cells increases protection against oxidative stress. Free Radical Biol Med. 2000; 29:51–60.

4. Thole JM, Kraft TF, Sueiro LA, Kang YH, Gills JJ, Cuendet M, et al. A comparative evaluation of the anticancer properties of European and American elderberry fruits. J Medicinal Foods. 2006; 9:498–504.

Does the Nutritional Content of Berries Vary Depending on Whether They Were Grown Organically or Conventionally?

By Michele Repo, MD

Berries have received no shortage of good press and are firmly established as a "healthy" food. But, are all berries created equal? Or are there factors that can make one bowl of berries more nutritious than another? Well, as a start, we can look at the work of researchers hoping to answer this question: "Does the nutritional content of berries that have been grown organically differ from those that have been grown conventionally?

What do the studies tell us so far? That's an important question but before delving into the answer(s), let's first establish what is being compared. When it comes to the nutritional content of berries, some of the molecules of particular interest are

phytonutrients (quite simply, plant nutrients). These chemicals are also referred to as "bioactive compounds" (or BAC for short). Why the interest in these BACs? Well, it's thought that they have certain functions (e.g. anti-inflammatory and anti-oxidant) that may prove beneficial to humans.

Naturally, if you're a health-conscious consumer, it becomes important to know which berries offer the most potential benefit. And, with an aging population and an impressive list of chronic diseases to address in North America, it is understandable that foods that offer the greatest nutritional benefit have caught the attention of scientists far and wide.

Despite health being an important driver of studies on berries, keep in mind that the studies reviewed in this report make no function or health-related claims. They are simply comparing levels of certain phytonutrients in organic and conventionally produced berries.

Photo Credit: Shutterstock

It would be great if all studies on nutrient levels in berries compared the same BACs. Unfortunately, that's not the case. For that reason, research on organic versus conventionally grown berries can't simply be thrown together to arrive at a single conclusion. Likewise, the studies don't always look at the same berries. As a particularly popular berry, research on strawberries currently far outweighs research on other berries such as blueberries, blackberries, and elderberries.

Rather than inundate you with a long list of numbers, perhaps a table (or three) will better illustrate some of the relevant research findings. And apologies for the long chemical names that sometimes surface in the tables. Such is the nature of BACs!

Let's start by looking at blueberries grown in New Jersey (Table 5). When they were compared, statistically significant differences in BAC content were found between organic and conventionally grown fruit. Based on this study, we can score one for organic berries.(1)

Table 5: Bioactive Compounds (BAC) of Organic Versus Conventional Blueberries Grown in New Jersey

SAMPLE	ANTI-OXIDANT ACTIVITY (MICROMOLES/ G OF FWT)	TOTAL ANTHOCYANINS (MG/100 G FWT)	TOTAL PHENOLICS (MG/100 G FWT)
Organic Blueberries	46.14	131.02	319
Conventionally grown blueberries	30.76	82.36	190

fwt = fresh weight

Moving on to two different types of strawberries grown in Maryland, significant differences in some BAC levels were found when comparing conventionally and organically grown fruit. Again, the organic berries came out ahead.(2)

Table 6: Bioactive Compounds (BAC) of Organic Versus Conventional Strawberries Grown in Maryland

CULTIVAR	ORGANIC/ CONVENTIONAL	BAC #1 (1)	BAC #2 (1)	BAC #3 (2)	BAC #4 (3)
#1	Organic	27.1	18.6*	12.6*	2.4
#1	Conventional	16.5	14.7*	8.0*	2.1
#2	Organic	34.7*	18.5	33.1	2.6*
#2	Conventional	25.8*	12.2	31.0	1.9*

(1) Data expressed as microgram of ellagic acid equivalents per gram of fresh weight.
(2) Data expressed as microgram of quercetin 3-glucoside equivalents per gram of fresh weight.
(3) Data expressed as microgram of kaempferol 3-glucoside per gram of fresh weight.
* statistically significant differences

Lastly, moving west to California, a study of organic strawberries showed that they too contained significantly higher levels of certain phytonutrients.(3)

Table 7: Phytonutrient Content of Organic Versus Conventional Strawberries Grown in California

SAMPLE	ANTI-OXIDANT ACTIVITY (1)	TOTAL PHENOLICS (2)	TOTAL ASCORBIC ACID (3)
Organic Strawberries	11.88	1.37	0.621
Conventionally grown strawberries	10.95	1.24	0.566

(1) mmol Trolox equivalents/grams fruit (2) mg gallic acid equivalents/gram fruit (3) mg/g fruit

This study also demonstrated that the organically grown strawberries had lower potassium and phosphorus levels compared to the conventionally grown berries.

Organic raspberries have also been found to offer higher amounts of antioxidant activity and flavonoids.(4) In the same vein, the phytonutrient level of both blueberries and blackberries has been found to be higher in wild (and therefore organic) versus conventionally grown crops.(5) Of course, research results are not uncommonly contradictory. So, it shouldn't come as any surprise that a 2011 study described the differences in phytonutrient levels between organically and conventionally produced blueberries to be "subtle" rather than significant.(6)

A 2012 review of organic versus conventional foods concluded that there was no strong evidence that the former is more nutritious than the latter. However, this review was not specific to either berries or BACs. The authors of the review also acknowledged that organic foods may reduce the risk of exposure to pesticide residue and antibiotic-resistant bacteria.(7) It's also important to recognize that organic versus conventional is only part of the BAC story. There are many other factors that can have an impact on phytonutrient levels in berries. These variables include time of year, maturity of the berries, cultivar and growing season. In other words, it may make sense to consider more than just growing method when looking for the berries with the best BAC scores.

Why Then the Difference

It seems fair to say that differences in phytonutrient content between organic and conventionally produced berries is not a one-off finding. This then begs the question of why does the cultivation method influence a berry's BAC level?

Several theories have been put forth including one based on how plants allocate resources.(8) With a finite supply of energy, plants must distribute it between competing functions, two important ones of which are growth and defense. When fungicides and/or insecticides are applied (as per conventional agriculture), the berries are provided an external source of protection. This presumably allows them to move some of their resources away from defense and more towards growth. How is that relevant to phytonutrients?

Well, phytonutrients play an important role in plant defense mechanisms – when the need for them is less, it makes sense that the plant decreases its production of, for example, antioxidants. Presto, you end up with less BACs. Add to this the rapid growth that is facilitated by the fertilizers and you end up with an even lower amount of BACs/100 g fruit. In other words, plants give up some of their BAC based protection to make the most of the opportunity to grow like stink in the presence of fertilizer. This may seem like a good deal to the plant, but may be less so for the consumer who wants berries with optimized phytonutrient content.

Can't I Just Wash My Berries - and Presto, Organic Berries Emerge?

Given the economic value of berries, it perhaps comes as no surprise that strawberries, blueberries, and currants have been described as "intensively chemically protected".(9) Also, not surprising is that some of this "protection" ends up on the fruit as chemical residue. Given the health consciousness of today's consumers, the possibility that they may be ingesting pesticides, even at levels considered non-toxic, often doesn't sit very well. As a result, consumers often want to know if certain berries pose more of a pesticide problem than others.

In the USA, and based on information from the EPA, strawberries have been classified by the Environmental Working Group as being the worst of the "dirty dozen," i.e., the twelve fruits and vegetables with the most pesticide residue. In comparison, domestic blueberries came in at #17, imported blueberries at #20 and raspberries at #24. In an attempt to contribute to the discussion about pesticide residue on berries, researchers from Poland had these results to offer with regards to a common fungicide, dithiocarbamate:(9)

Table 8: Fungicide Residue on Berries Grown in Poland

BERRY	% CONTAMINATED WITH DITHIOCARBAMATE	TOTAL ANTHOCYANINS (MG/100 G FWT)
Strawberries	6.1	.09 – 1.49
Red Currant	8.3	.08 – .58
Raspberries	17.2	.09 – .24
Gooseberry	27.7	.06 – .28
Black Currant	28.6	.05 – .10

Having established that pesticide residue is in fact a real possibility in berry crops, what then do we know about washing said berries? As a hotbed of berry production, researchers from Poland again have something to add to the discussion.

In a 2016 report, researchers demonstrated that five minutes of washing berries with water removed more residue than washing for one or two minutes. That was the good news. The bad news was that even after five minutes, pesticide residue could still be detected on the strawberries.(10) (Further, one wonders, how many consumers wash their berries for a full five minutes?). This finding was supported by an earlier study out of Italy which showed that both washing with water and washing with detergent reduced but did not eliminate pesticide residue on strawberries.(11) Closer to home, in a study of low bush blueberries in Michigan, washing them with distilled water reduced the amount of phosmet, an organophosphate insecticide, found on the fruit. (12) Although the post-washing level of phosmet was below the level considered safe by the EPA, this may not be a reassuring finding for those people who consider any amount of pesticide a cause for concern.

This is particularly so because fungicides are one of the most likely contaminants of conventional berries and, given their adverse health effects, are also particularly unwelcome guests.(9) Lastly, washing does nothing to address any differences in nutrient content between organically and conventionally grown berries.

Any Other Factors to Consider Before I Buy a Pint of Elderberries (or Blueberries, or Blackberries)?

Just as there is no shortage of interest in berries and their nutrients, there is also no shortage of interest in urban farming. There are many factors driving this interest, including food security, finding a use for marginal or abandoned land, and helping city dwellers forge a closer connection to the food they eat.

But are berries grown in the city nutritionally on par with berries grown in the country? Answering that question definitively will require more work but research out of, (you guessed it!), Poland is certainly relevant to this discussion. A 2009 study in the south-eastern part of that country looked at the flavonoid (phytonutrient) content of elderberries in relation to where they were collected.

Berries picked at sites that were closer to high traffic transportation routes had lower levels of flavonoids (bad), and higher levels of heavy metals such as lead, chromium, cadmium and copper (also bad).(13) An earlier study in Denmark found that when grown in contaminated soil, elderberries had higher levels of cadmium than their grocery store counterparts. Similarly, berries that grow in bushes (as opposed to trees) had higher levels of lead contamination – presumably due to dust drifting up from the contaminated soil.(14)

When it comes to the impact of traffic on the BAC levels found in berries, North American research is not easy to come by. What is more readily available are multiple studies that document increases in heavy metals such as lead and cadmium in soil plots closer to major roads. What this means is that even if plants don't incorporate these heavy metals into their berries, surface contamination can be a concern in those kinds of plots.(15, 16)

And so, the "health" of berries grown in urban settings may be significantly influenced by the quality of the soil, which in turn is influenced by the amount of local traffic pollution, amongst other factors. Okay, understood.

But is soil contamination in urban gardens a big problem or a once-in-a-lifetime situation? Well, a study published in 2014 demonstrated that of 55 community gardens in New York City, 70% had at least one soil sample with higher than recommended levels of metals. Leading the pack were the heavy metals, barium and lead, which exceeded healthy levels in 12% and 9% of samples, respectively.(17)

There are many benefits to urban gardening but it is imperative that soil safety be confirmed. With these results in mind, "location, location, location", may be as relevant to growing berries as it is to real estate!

Summary

Apparently when it comes to nutrient content, a berry is not necessarily a berry. There are many factors that can influence nutrient content, not the least of which is whether organic or conventional growing methods have been used. It would also be nice to know whether the berries have been grown far enough away from traffic pollution to be spared some of its heavy metals.

Although not providing a sure-fire guarantee that you're about to eat the most nutritious berry possible, the literature does provide some evidence that you may get more phytonutrient bang for your berry buck when you opt for the organically grown version. But, if at all possible, make sure your organic berries aren't grown near a major roadway or in soil that has become contaminated with toxins in another way.

Otherwise, the luster of organic berries may lose some of its sheen.

REFERENCES

1. Wang, S.Y., Chen, C.T., Sciarappa, W., Wang, C.Y., & Camp, M.J. (2008). Fruit quality, antioxidant capacity, and flavonoid content of organically and conventionally grown blueberries. J AgricFood Chem, 56(14), 5788-94.

2. Jin, P., Wang, S.Y., Wang C.Y., & Zheng, Y. (2011). Effect of cultural system and storage temperature on antioxidant capacity and phenolic compounds in strawberries. Food Chem, 124, 262–270. doi: 10.1016/j.foodchem.2010.06.029.

3. Reganold, J.P., Andrews, P.K., Reeve, J.R., Carpenter-Boggs, L., & Schadt C.W. (2015). Fruit and Soil Quality of Organic and Conventional Strawberry Agroecosystems. PLoS ONE, 5, 1-14. doi: 10.1371/annotation/1eefd0a4-77af-4f48- 98c3-2c5696ca9e7a.

4. Jin, P., Wang, S.Y., Gao, H., Chen H., Zheng, Y., & Wang C.Y. (2012). Effect of cultural system and essential oil treatment on antioxidant capacity in raspberries. Food Chem, 132, 399–405. doi: 10.1016/j.foodchem.2011.11.011.

5. Koca, I., & Karadeniz, B. (2009). Antioxidant properties of blackberry and blueberry fruits grown in the Black Sea Region of Turkey. Scientia Horticulturae, 121(4), 447-450.

6. You, Q., Wang, B., Chen, F., Huang, Z., Wang, X., & Luo, P. G. (2011). Comparison of anthocyanins and phenolics in organically and conventionally grown blueberries in selected cultivars. Food Chemistry, 125(1), 201-208, http://dx.doi.org/10.1016/j.foodchem.2010.08.063

7. Smith-Spangler, C., Brandeau, M.L., Hunter, G.E., Bavinger, J.C., Pearson, M., Eschbach, P.J., Sundaram, V., Liu, H., Schirmer, P., Stave, C., Olkin, I., & Bravata, D.M. (2012). Are Organic Foods Safer or Healthier Than Conventional Alternatives?: A Systematic Review. Ann Int Med, 157(5), 348-366. DOI: 10.7326/0003-4819-157-5- 201209040-00007

8. Stamp, N.E. (2003). Out of the quagmire of plant defense hypotheses. Q. ReV. Biol, 78, 23-55.

9. Wolejko, E., Kozowicka, B., & Kaczyński, P. (2014). Pesticide residues in berries fruits and juices and the potential risk for consumers. Desalination and Water Treatment, 52(19-21), 3804-3818, DOI:10.1080/19443994.2014.883793

10. Lozowicka, B., Rutkowska, E., Jankowska, M., Kaczyński, P., & Hyrnko, I. (2012). Health risk analysis of pesticide residues in berry fruit from north-eastern Poland. J Fruit Orna Plant Res, 20(1), 83-95.

11. Angioni, A., Schirra, M., Garau, V.L., Melis, M., Tuberoso, C.I.G., & Cabra, P. (2004) Residues of azoxystrobin, fenhaxamid and pyrimethanil in strawberry following field treatments and the effects of domestic washing. Food Addit Contam, 21(11), 1065-70.

12. Crowe-White, K.M. (2002). Effects of Post-harvest treatments on the microbiological quality and pesticide residues of Lowbush Blueberries. (Masters Dissertation). University of Michigan.

13. Kołodziej, B., Maksymiec, N., Drożdżal, K., & Antonkiewicz, J. (2012). Effect of traffic pollution on chemical composition of raw elderberry (Sambucus nigra L.). Journal of Elementology, 17(1) 67-78.

14. Samsøe-Petersen, L., Larsen, E.H., Larsen, P.B., & Bruun, P. (2002). Uptake of Trace Elements and PAHs by Fruit and Vegetables from Contaminated Soils. Environ Sci Technol, 36(14), 3057-63.

15. McBride, M.B. Shayler, H.A., Spliethoff, H.M., Mitchell, R.G., Marquez-Bravo, L.G., Ferenz, G.S., Russell-Anelli, J.M., Casey, L., & Bachman, S. (2014) Concentrations of lead, cadmium and barium in urban garden-grown vegetables: the impact of soil variables. Environ Pollut, 194, 254-61. doi: 10.1016/j.envpol.2014.07.036. Epub 2014 Aug 28.

16. Clarke, L.W., Jenerette, G.D., & Bain, D.J. (2015). G. Urban legacies and soil management affect the concentration and speciation of trace metals in Los Angeles community garden soils Environ Pollut, 197, 1-12.

17. Mitchell, R.G., Spliethoff, H.M., Ribaudo, L.N., Lopp, D.M., Shayler, H.A., Marquez-Bravo, L.G., Lambert, V.T., Ferenz, G.S., Russell-Anelli, J.M., Stone, E.B., & McBride,

18. M.B. (2014). Lead (Pb) and other metals in New York City community garden soils: factors influencing contaminant distributions. Environ Pollut, 187, 162-9. doi: 10.1016/j.envpol.2014.01.007. Epub 2014 Feb 3

How Good a Source of Polyphenols is Elderflower Tea?

By Susan Elrod, Phd

Various fruits, herbs, and flowers are used to make herbal teas. These herbal infusions (commonly called teas, though they contain no actual Camellia sinensis, the plant used to make green, black, or white tea) are typically blended for flavor but also may be used for medicinal benefits.

Lavender and chamomile, for example, are commonly associated with relaxation, as are ginger and licorice with stomach relief. Elderberry and elderflower are both often included in herbal teas. Other chapters in this book have discussed the phenolic content of elderberry, but what about elderflower?

Elderflower, like many flowers, is edible and often used in beverages. In addition to herbal tea, you can find it in sodas, wines, beer, and (perhaps most famously) St. Germaine liqueur. If you're a fan of the light, sweet flavor of these flowers, there's good news: elderflower tea may have a higher phenolic content than elderberry tea(1).

Regardless of what part of the elder tree comes out on top in terms of phenolic content, elderflower does contain significant amounts of polyphenols.

As you might expect, however, elderflowers have different types of polyphenols compared to elderberries. As discussed in other articles, the berries have a great deal of cyanidin-related compounds, as well as several other polyphenols. Since the cyanidins give elderberries that dark purple color, it is unsurprising that they are absent from the flowers, However, the flowers do seem to have a great deal of other compounds that have been studied and associated with medicinal benefit, including quercetin, kaempferol, and naringenin.

In a study of the effects of elderberry and elderflower extracts on markers of inflammation, elderflower performed at least as well and often better than elderberry (that is, it took lower concentrations of the elderflower extract to yield the same effect compared to elderberry).(2)

What about when these flowers and berries are used specifically for teas? There is good news for elderflower fans there as well. The study conducted on the phenolic content of elderberry and elderflower was performed using infusions, that is, by adding boiling water to the berries and flowers in a manner similar to that used to prepare herbal teas.(1)

That means that the results of the study indicate the amount of polyphenols present in such herbal teas, so you don't have to worry about heat destroying or compromising the phenols in elderflower tea. Furthermore, in terms of polyphenol content, elderberry and elderflower seem to compare well to other herbal tea ingredients, on par with chamomile and potentially higher than ginger or turmeric*.(1, 3-5)

Elderflower tea was also comparable to green and white tea in a study of chlorogenic acid content (studies suggest chlorogenic acid is responsible for many benefits associated with coffee consumption). (6) The exact amount of polyphenols in any herbal extract will vary according to preparation as well as the plant species, cultivar, and growing method, but it appears that elderberry and elderflower tea both provide significant amounts of polyphenols.

Photo Credit: Shutterstock

*Based on comparison of results of different studies and conversion of units used for said studies. Side by side comparison may provide different results.

REFERENCES

1. Viapiana A, Wesolowski M. The phenolic and antioxidant activities of infusions of Sambucus nigra L. Plant Foods Hum Nutr 2017;72(1):82-87.

2. Ho GTT, Wangensteen H, Barsett H. Elderberry and elderflower extracts, phenolic compounds, and metabolites and their effect on complement, RAW 264.7 macrophages and dendritic cells. Int J Mol Sci 2017;18(3):584.

3. Maizura M, Aminah A, Wan Aida W. Total phenolic content and antioxidant activity of kesum (Polygonum minus), ginger (Zingiber officinale) and turmeric (Curcuma longa) extract. International Food Research Journal 2011;18(2).

4. Kaur C, Kapoor HC. Anti oxidant activity and total phenolic content of some Asian vegetables. Int J Food Sci Tech 2002;37(2):153-161.

5. Park E, Bae W, Eom S, Kim K, Paik H. Improved antioxidative and cytotoxic activities of chamomile (Matricaria chamomilla) florets fermented by Lactobacillus plantarum KCCM 11613P. Journal of Zhejiang University-SCIENCE B 2017;18(2):152-160.

6. Meinhart AD, Damin FM, Caldeirão L, da Silveira TFF, Filho JT, Godoy HT. Chlorogenic acid isomer contents in 100 plants commercialized in Brazil. Food Research International 2017 September 2017;99(Part 1):522-530.

Maximizing the Antioxidant Power of Berries

By Melissa M. Melough, RD

Berries contain many health-promoting nutrients including vitamins, minerals, and fiber. One reason they're often heralded as "superfoods" is their high content of antioxidants.

Antioxidants are compounds that prevent damage caused by the oxidation of important structures such as membranes, lipids, proteins, and DNA. Antioxidants are found in a variety of foods, especially fruits and vegetables. Berries are particularly rich in antioxidants because they contain several different antioxidants including vitamin C, vitamin E, selenium, and many phenolic compounds, a class of compounds produced by plants. In fact, an analysis of over 3,100 food samples showed that berries and berry products were the most antioxidant-rich category of foods except for supplements and spices.(1)

Anthocyanins are one key group of phenolic compounds in berries that act as antioxidants. Anthocyanins are pigments that give berries their red, purple, or blue color, and they account for roughly 45% of the total antioxidant activity of many berries.(2) Elderberries are particularly high in anthocyanins – which give them their purple color – and are one example of an antioxidant-rich berry. Comparing levels of antioxidants across different berries is challenging because of natural variation, but based on the measurements of many samples of fresh berries by one research team, elderberries have greater antioxidant activity than many common berries such as blueberries, cranberries, raspberries, and strawberries.(1)

Table 9: Levels of Antioxidant Content of Different Types of Berries

BERRY	ANTIOXIDANT CONTENT (MMOL/100G)*
Bilberries	8.1
Chokeberries	8.0
Blackcurrant	7.3
Blackberries	4.5
Elderberries	4.3
Blueberries	3.6
Cranberries	3.3
Cloudberries	2.9
Raspberries	2.8
Strawberries	2.1
Redcurrant	1.6

Adapted from data presented by Carlsen et al in the Nutrition Journal (2010)(1) using averaged measurements of fresh berry samples

The exact level of antioxidants in a berry is affected by many factors out of consumers' control such as growing location and nutrients in the soil. However, there are several important factors impacting the level of antioxidants that consumers can control. First, storage impacts some antioxidants. One study examined the anthocyanin levels in blueberries stored for up to 3 weeks. The anthocyanin levels of some blueberry species actually increased over time when stored at room temperature or in the refrigerator or freezer.(3)

After longer periods of time, though, anthocyanin content is likely to decline. In one study, the level of anthocyanins in elderberry juice was significantly lower at 3 months of frozen storage than when it was originally produced.(4) Other important antioxidants like vitamin C can also degrade over time in fruits and juices.(5) Therefore, consuming berries or berry products shortly after purchase is likely the best way to maximize the antioxidant benefit. Antioxidants in fruits such as berries can also be affected by processing. One study showed that antioxidant activity was decreased by about 90% in black mulberries made into jam using an industrial process involving relatively high heat.(6)

However, this study and others also showed that the processing methods used to make jams and juices actually increase the bioavailability of the antioxidants.(6–8) Therefore, heating can reduce the level of antioxidants, but make it easier for the body to absorb them. This means that berry jams, syrups, and juices can still be a good way to get antioxidants.

Research shows that lighter cooking methods – such as baking for 20 minutes at 375°F or briefly microwaving or pan-frying – does not reduce the antioxidant content of blueberries.(9) Baking raspberry juice into muffin batter reduced the amount of anthocyanins, but overall it increased the total level of antioxidants.(10) Another study showed that freezing or drying blueberries did not change the level of antioxidants or anthocyanins compared to the fresh fruit.(11) Each of these studies shows that light processing such as freezing or gently heating does not reduce the antioxidant benefit of berries.

Photo Credit: Shutterstock

People often wonder if the antioxidant level of berries can be affected by the other foods they eat along with the berries. In the early 2000's a couple of small studies led to sensational headlines making people fear that milk could "block" antioxidants. (12,13) These studies suggested that milk could reduce the absorption of antioxidants from blueberries(13) and chocolate,(12) but they were both conducted by the same group of researchers, and included data from only 11 or 12 volunteers.

Therefore, this evidence is rather weak. In fact, another small study drew the opposite conclusion: drinking skim milk along with blackberry juice had no significant impact on antioxidant absorption.(14) Therefore, more research on this topic is needed. For now, it's probably best to continue eating antioxidant-rich berries in a variety of ways, both on their own and mixed in with other foods.

The bottom line is that berries are one of the best sources of antioxidants, and elderberries rank among the top berries for antioxidant content. Eating fresh or minimally processed berries soon after purchasing them is a great way to maximize your dose of antioxidants. However, eating berry products like jams, juices, and syrups may actually make it easier to absorb the antioxidants, so including these in your diet can be a great source of antioxidants too.

REFERENCES

1. Carlsen, M. H. et al. The total antioxidant content of more than 3100 foods, beverages, spices, herbs and supplements used worldwide. Nutr. J. 9, 3 (2010).

2. Lee, S. G. et al. Contribution of Anthocyanin Composition to Total Antioxidant Capacity of Berries. Plant Foods Hum. Nutr. 70, 427–432 (2015).

3. Mallik, A. U. & Hamilton, J. Harvest date and storage effect on fruit size, phenolic content and antioxidant capacity of wild blueberries of NW Ontario, Canada. J. Food Sci. Technol. 54, 1545–1554 (2017).

4. Johnson, M. C., Thomas, A. L. & Greenlief, C. M. Impact of Frozen Storage on the Anthocyanin and Polyphenol Contents of American Elderberry Fruit Juice. J. Agric. Food Chem. 63, 5653–5659 (2015).

5. Castro-López, C. et al. Fluctuations in phenolic content, ascorbic acid and total carotenoids and antioxidant activity of fruit beverages during storage. Heliyon 2, (2016).

6. Tomas, M. et al. Processing black mulberry into jam: effects on antioxidant potential and in vitro bioaccessibility. J. Sci. Food Agric. 97, 3106–3113 (2017).

7. Tomas, M. et al. The effects of juice processing on black mulberry antioxidants. Food Chem. 186, 277–284 (2015).

8. Toydemir, G. et al. Changes in sour cherry (Prunus cerasus L.) antioxidants during nectar processing and in vitro gastrointestinal digestion. J. Funct. Foods 5, 1402–1413 (2013).

9. Murphy, R. R., Renfroe, M. H., Brevard, P. B., Lee, R. E. & Gloeckner, J. W. Cooking does not decrease hydrophilic antioxidant capacity of wild blueberries. Int. J. Food Sci. Nutr. 60, 88–98 (2009).

10. Rosales-Soto, M. U., Powers, J. R. & Alldredge, J. R. Effect of mixing time, freeze-drying and baking on phenolics, anthocyanins and antioxidant capacity of raspberry juice during processing of muffins. J. Sci. Food Agric. 92, 1511–1518 (2012).

11. Lohachoompol, V., Srzednicki, G. & Craske, J. The change of total anthocyanins in blueberries and their antioxidant effect after drying and freezing. J. Biomed. Biotechnol. 2004, 248–252 (2004).

12. Serafini, M. et al. Plasma antioxidants from chocolate. Nature 424, 1013 (2003).

13. Serafini, M. et al. Antioxidant activity of blueberry fruit is impaired by association with milk. Free Radic. Biol. Med. 46, 769–774 (2009).

14. Hassimotto, N. M. A., Pinto, M. D. S. & Lajolo, F. M. Antioxidant status in humans after consumption of blackberry (Rubus fruticosus L.) juices with and without defatted milk. J. Agric. Food Chem. 56, 11727–11733 (2008).

Section 2: What's the Best Way to Prepare Elderberries to Eat?

Are Elderberries Safe to Eat: Raw, Heated or Otherwise Processed?

By Eugene Fenster, PhD

Where elderberries are grown and how they are prepared have an impact on how safe they are to consume and the amount of nutrients, and non-nutrients, the body obtains from them. If elderberry is grown in an area that is polluted, some of the pollutants may be incorporated into the plant. If elderberry is subjected to heat, there will be a loss of organic compounds, yet on the other hand if it is not subjected to sufficient heat then contaminating pathogens will not be killed.

The Importance of Location

A study that took place in Poland considered how distance to traffic and the amount of traffic affect the mineral and phytochemical content of elderberries.(1) It was found that the greater the road traffic and the closer to traffic the lower the concentration of phytochemicals and the higher the concentrations of chromium, iron, cadmium, and copper. While chromium, iron, and copper are essential nutrients, cadmium is detrimental to health. Even at low levels of intake cadmium may be toxic and have an adverse impact on liver and kidney function, can weaken bones, and raise blood pressure.(2) Studies of mice have shown that the consumption of cadmium can have an adverse impact on the ovaries and testes and may increase the risk of kidney cancer. Prenatal exposure to cadmium may increase the risk of having low birth weight babies as well as the risk of spontaneous abortion(3).

Consequences of Preparation Method

The longer a food, including elderberries, is subject to heat, and the higher the heat, the greater the loss of organic nutrient and phytochemical content. The result is a reduction of the health benefits conferred by the consumption of these foods.(4) For example, elderberries can be spray dried (a method of food preservation) with a variety of substances and across a range of temperatures.

Regardless of the spray drying method, there is a loss of phytochemicals(5). The production of fruit spreads requires heating of the fruit to a temperature that kills potential pathogens. This same heating that makes the fruit spread safe to eat also reduces the phytochemical content of the spread.(6) Even processing such as making a juice can reduce phytochemical content.(7) Despite this loss of phytochemicals with juicing, both raw elderberries and elderberry juice are considered a rich source of phytochemicals.(8)

A small sample size study found that a high-fat diet promotes the absorption of the phytochemicals found in elderberries. Then, within a few hours of their consumption, the amounts of some of the phytochemicals in the blood was less than 1% of the amount that was consumed.(9) Many phytochemicals are not readily retained in the body.

The retention of these phytochemicals has been seen to be enhanced by consuming them along with sucrose.(10,11) However, adding table sugar, sucrose, to berries to enhance phytochemical retention is counterproductive given the adverse health consequences of added sugar consumption.

Potential Safety Issues

Not all of the substances found in elderberries promote health. Elderberries, particularly the bark, leaves, seeds, and unripe fruit, contain compounds that have the potential to release cyanide.(12,13) Processing will reduce the amounts of these cyanide-releasing compounds. Regardless, it may be wise to avoid consuming the bark, leaves, roots, and unripe berries because of the risk of cyanide poisoning. Short-term drinking of tea made from elderberry flowers has been found to be safe. (6) Symptoms of cyanide poisoning include dizziness, nausea, vomiting, difficulty breathing, and a rapid heart rate.(14)

In 1983, 11 cases of elderberry poisoning (from a species different from the one grown at Fat Stone Farm) were reported.(15) The reported symptoms were nausea, vomiting, abdominal cramps, and weakness. Some also complained of dizziness and numbness, and one person was nearly unconscious and had to be hospitalized. These individuals were among a group of 25 people who had gathered wild elderberry and prepared a juice by crushing the berries, leaves, and branches in a stainless-steel press. They then added apple juice, water, and sugar and stored the mixture overnight. The severity of the symptoms increased with the amount of juice consumed.

People with allergies, especially those with allergies to plants of the honeysuckle family, may have a reaction to elderberry pollen and elderberry.(6) Exposure to elderberry pollen or the consumption of elderberry-containing products may trigger a reaction in those who suffer from allergic rhinitis.(16)

A laboratory study that simulated digestion of elderberries concluded that boiling elderberries for 5 – 10 minutes prior to consumption reduced/eliminated the allergic response as it enabled the body to digest the allergens. However, doing so also reduced the antioxidant potential of the elderberry by a bit under 10%.(17)

Recommendations from the USDA(18)

The USDA considers raw (fresh) purple or blue elderberries to be safe to consume, in moderation. They also state that purple or blue elderberries can be used to make elderberry wine, jam, syrup, and pies. In addition, they state that the entire flower cluster can be dipped in batter and fried, and the petals can be eaten raw or made into a tea or added to pancakes.

While this may seem to contradict what was written earlier, it is important to realize that moderation is key and that most of these suggestions involve the use of heat which reduces the amount of harmful organic substances present in this, and every, food.

Conclusions

In summation, enjoy washed raw blue or purple elderberries, in moderation; enjoy elderberry fruit spreads, in moderation; enjoy dried elderberry fruit, in moderation. Doing so will be of benefit to your health. Have an infrequent cup of tea made from elderberry flowers. Avoid consuming elderberry bark, leaves, seeds, and unripe fruits or products made from them.

REFERENCES

1. Kołodziej, B., Maksymiec, N., Drożdżal, K., & Antonkiewicz, J. EFFECT OF TRAFFIC POLLUTION ON CHEMICAL COMPOSITION OF RAW ELDERBERRY (SAMBUCUS NIGRA L.).

2. Duruibe, J. O., Ogwuegbu, M. O. C., & Egwurugwu, J. N. (2007). Heavy metal pollution and human biotoxic effects. International Journal of Physical Sciences, 2(5), 112-118.

3. Godt, J., Scheidig, F., Grosse-Siestrup, C., Esche, V., Brandenburg, P., Reich, A., & Groneberg, D. A. (2006). The toxicity of cadmium and resulting hazards for human health. Journal of occupational medicine and toxicology, 1(1), 22.

4. Patras, A., Brunton, N. P., O'Donnell, C., & Tiwari, B. K. (2010). Effect of thermal processing on anthocyanin stability in foods; mechanisms and kinetics of degradation. Trends in Food Science & Technology, 21(1), 3-11.

5. Murugesan, R., & Orsat, V. (2011). Spray drying of elderberry (Sambucus nigra L.) juice to maintain its phenolic content. Drying Technology, 29(14), 1729-1740.

6. Cejpek, K., Maloušková, I., Konečný, M., & Velíšek, J. (2009). Antioxidant activity in variously prepared elderberry foods and supplements. Czech J. Food Sci, 27, 45-48.

7. Casati, C. B., Sánchez, V., Baeza, R., Magnani, N., Evelson, P., & Zamora, M. C. (2012). Relationships between colour parameters, phenolic content and sensory changes of processed blueberry, elderberry and blackcurrant commercial juices. International journal of food science & technology, 47(8), 1728-1736.

8. Galić, A., Dragović-Uzelac, V., Levaj, B., Bursać Kovačević, D., Pliestić, S., & Arnautović, S. (2009). The polyphenols stability, enzyme activity and physico-chemical parameters during producing wild elderberry concentrated juice. Agriculturae Conspectus Scientificus (ACS), 74(3), 181-186.

9. Murkovic, M., Toplak, H., Adam, U., & Pfannhauser, W. (2000). Analysis of anthocyanins in plasma for determination of their bioavailability. Journal of Food Composition and Analysis, 13(4), 291-296.

10. Mülleder, U., Murkovic, M., & Pfannhauser, W. (2002). Urinary excretion of cyanidin glycosides. Journal of biochemical and biophysical methods, 53(1), 61-66.

11. Kaşıkcı, M. B., & Bağdatlıoğlu, N. (2016). Bioavailability of Quercetin. Current Research in Nutrition and Food Science Journal, 4(Special Issue Nutrition in Conference October 2016), 146-151.

12. Ulbricht, C., Basch, E., Cheung, L., Goldberg, H., Hammerness, P., Isaac, R., & Purkh, K. (2014). An Evidence-Based Systematic Review of Elderberry and Elderflower (Sambucus nigra) by the Natural Standard Research Collaboration. Journal of Dietary Supplements, 11(1), 80-120.

13. Senica, M., Stampar, F., Veberic, R., & Mikulic-Petkovsek, M. (2016). Processed elderberry (Sambucus nigra L.) products: A beneficial or harmful food alternative?. LWT-Food Science and Technology, 72, 182-188.

14. CDC. Cyanide. https://emergency.cdc.gov/agent/cyanide/basics/facts.asp. Last update: November 18, 2015. Accessed: July 19, 2017

15. MMWR Weekly. 1984. Poisoning from Elderberry Juice – California. 33(13):173-174

16. Förster-Waldl, E., Marchetti, M., Schöll, I., Focke, M., Radauer, C., Kinaciyan, T., Nentwich, I., Jäger, S., Schmid, E.R., Boltz-Nitulescu, G., Scheiner, O. and Jensen-Jarolim, E. (2003), Type I allergy to elderberry (Sambucus nigra) is elicited by a 33.2 kDa allergen with significant homology to ribosomal inactivating proteins. Clinical & Experimental Allergy, 33: 1703–1710. doi:10.1111/j.1365-2222.2003.01811.x

17. Jiménez, P., Cabrero, P., Cordoba-Diaz, D., Cordoba-Diaz, M., Garrosa, M., & Girbés, T. (2017). Lectin Digestibility and Stability of Elderberry Antioxidants to Heat Treatment In Vitro. Molecules, 22(1), 95.

18. Common elderberry-USDA plants database. USDA Plant Guide. pp. 1-6. 35.

Elderberry Skins: Are They Important?

By Susan Elrod, PhD

You may have heard that fruit skins contain some of the best nutrients. Much of this is fiber, but other nutrients are present in fruit and vegetable skins as well.

In some cases this is interesting but irrelevant information; you wouldn't want to eat your orange peel, no matter how many more polyphenols and other nutrients you'd be getting.(1) But what about fruits and vegetables with edible skins? What does the evidence say about consuming the whole fruit, skin and all, as opposed to using juices or extracts?

First, let's review the definition of some commonly used terms for purposes of our scientific discussion. A whole fruit (a single berry or pomme fruit, for example) is comprised of skin (epicarp, or the outer edge of the fruit), mesocarp (which contains both flesh and juice inside), endocarp (a membrane inside that covers the seeds) and the seed.

In scientific experiments, we can separate these different parts, but in everyday life, such as in juicing, you would be separating the juice from everything else (called the mash). One additional term that you will hear is pulp. Pulp is what remains after the skin is removed.

So what is the evidence about elderberries? There haven't been studies looking at the skin versus fruit (or "pulp," the term most frequently used in literature to distinguish the skin from the whole fruit). However, multiple studies looking at other berries and fruits have found that in general, skins tend to be higher in fiber and polyphenols. (2,3) Like a lot of food research, this varies according to circumstances; polyphenols and other compounds vary according to the environment and conditions of growing (4,5) and according to the fruit itself. A study of jamun (also known as Java plum), jackfruit, and mulberry found whole mulberry fruit was the highest in resveratrol (that compound so often cited for red wine's benefits), whereas Janum seeds had the highest phenolic and antioxidant content.(6)

That's not unusual, either; often seeds and skins of fruit are cited as excellent but often discarded sources of phenols and other nutrients. A study of grapes found that, while the skins had higher phenolic and antioxidant content than the pulp, the seeds were higher than both skin and pulp in these concentrations. (7) On the other hand, a study of chilto (or "tree tomato") found the skin to have the highest phenolic content, over pulp and seeds, whereas the pulp had the highest vitamin C content.(1,8)

A study of Amazonian berry found that not only did skins and seeds have a much higher phenolic content compared to pulp, but the specific polyphenols varied according to fruit part: both pulp and seeds/skins parts had myricetin and ellagic acid, but cyanidin and quercetin were found only in the pulp, and proanthocyanidins found only in the seeds/skin portion.

So does that mean you should start consuming fruits whole, seeds, skin and all? Not necessarily. When we report phenolic and antioxidant content, it's done on the basis of amount of polyphenols per weight (or volume) of fruit. So if you're told that the phenolic content of fruit skin is 1 mg/g, and the content for fruit pulp is 0.5 mg/g, you might believe that you'd be getting twice as much phenolic content by eating the skin.

However, consider the amount of skin as a portion of the whole fruit: it's a very small percentage. So if we assume that a 100 g fruit contains 90% pulp and 10% skin, the skin would contain only 10 mg polyphenols compared to 45 mg in the pulp.(1) Furthermore, a large volume of fruit (or pulp) goes into preparing juices, and the juice product will be more concentrated than the constituent fruits, typically yielding a higher concentration of polyphenols and other nutrients per volume.(9)

Leaving aside considerations of what part of the fruit generates what and how many polyphenols, we can also consider research into health benefits of fruits. Particularly in human studies, it's important to know what form the fruit was given to the patients. That is, was the study on an apple a day? A glass of apple juice a day? A dried apple preparation in a capsule a day? Looking at the original research will tell you what form of fruit was used to elicit the benefits seen by researchers.

Elderberry consumption has been studied in humans for benefit in a few different conditions, including cardiovascular disease. In such studies, elderberry juice and extract (a preparation of fruit made by passing a solvent through the dried fruit

to separate out the active ingredients) appeared to have no effect on improving serum lipids or biomarkers of cardiovascular disease.(9,10) Another study found an herbal tea made of four different plants, including elderberry, was efficacious against chronic constipation.(11)

Photo Credit: Shutterstock

The bulk of human research on elderberries, however, is on its effect against viruses, specifically the cold and flu. Although more research has been conducted on the effect of elderberry against viruses in vitro (that is, in cells, not in animals or humans), there is some limited evidence of elderberry's effect on viruses in human patients.

These studies all involve commercial preparations of elderberry. In the influenza studies, adults with influenza A or B taking four tablespoons of the elderberry extract daily for 3 or 5 days showed improvement 2 to 4 days sooner than those taking a placebo, depending on the study.(12,13)

In a study regarding cold symptoms, air travelers were given two capsules containing 300 mg of elderberry extract for ten days before their trip and three capsules daily during and up to 5 days after travel. Travelers taking the elderberry extract reported significantly shorter and less severe cold symptoms compared to the placebo group. (14)

So in summary, if you can consume whole fruits with the skin, that's great but may not be strictly necessary, given the weight by which whole fruits are consumed compared to juices. For elderberries specifically, the bulk of the research has been conducted using extracts, so the juices and concentrates should be just fine in terms of imparting any potential benefit.

REFERENCES

1. Antioxidant capacity and mineral content of pulp and peel from commercial cultivars of citrus from Brazil. Barros, Helena Rudge de Moraes, Ferreira, Tania Aparecida Pinto de Castro and Genovese, Maria Ines. 4, 2012, Food Chemistry, Vol. 134, pp. 1892- 1898.

2. Chemical characterization of the jabuticaba fruits (Myrciaria cauliflora Berg) and their fractions. Baori, Lime Ade J, et al. 4, 2008, Archivos Latinoamericanos de Nutricion, Vol. 58, pp. 416-421.

3. Inhibition of low-density lipoprotein oxidation by Nagano purple grape (Vitis viniferaxVitis labrusca). Kamiyama, Masumi, et al. 6, 2009, J Nutr Sci Vitaminol (Tokyo), Vol. 55, pp. 471-478.

4. Irrigation and Rootstock Effects on the Phenolic Concentration and Aroma Potential of Vitis vinifera L. cv. Cabernet Sauvignon Grapes. Koundouras, Stefanos, et al. 17, 2009, J. Agric. Food Chem., Vol. 57, pp. 7805-7813.

5. Determination of Anthocyanins and Total Polyphenols in a Variety of Elderberry Juices by UPLC-MS/MS and Other Methods. Wu, H, et al. 2015, Acta Hortic, Vol. 1061, pp. 43-51.

6. Resveratrol content and antioxidant properties of underutilized fruits. Shrikanta, Akshatha, Kumar, Anbarasu and Govindaswamy, Vijayalakshmi. 1, 2015, J Food Sci Technol, Vol. 52, pp. 383-390.

7. Identification and quantification of phenolic compounds in berry skin, pulp, and seeds in 13 grapevine varieties grown in Serbia. Pantelić, Milica M., et al. 2016, Food Chemistry, Vol. 211, pp. 243-252.

8. Chemical and functional characterization of seed, pulp and skin powder from chilto (Solanum betaceum), an Argentine native fruit. Phenolic fractions affect key enzymes involved in metabolic syndrome and oxidative stress. Orqueda, Maria Eugenia, et al. 2017, Food Chemistry, Vol. 216, pp. 70-79.

9. Effects of elderberry juice on fasting and postprandial serum lipids and low-density lipoprotein oxidation in healthy volunteers: a randomized, double-blind, placebo- controlled study. Murkovic, M, et al. 2004, European Journal of Clinical Nutrition, Vol. 58, pp. 244-249.

10. Cardiovascular Disease Risk Biomarkers and Liver and Kidney Function Are Not Altered in Postmenopausal Women after Ingesting an Elderberry Extract Rich

in Anthocyanins for 12 Weeks. Curtis, Peter J, et al. 12, 2009, The Journal of Nutrition, Vol. 139, pp. 2266-2271.

11. Randomized clinical trial of a phytotherapeutic compound containing Pimpinella anisum, Foeniculum vulgare, Sambucus nigra, and Cassia augustifolia for chronic constipation. Picon, Paulo D, et al. 17, 2010, BMC Complementary and Alternative Medicine, Vol. 10, pp. 10-17.

12. Inhibition of several strains of influenza virus in vitro and reduction of symptoms by an elderberry extract (Sambucus nigra L.) during an outbreak of influenza B Panama. Zakay-Rones, Zichria, et al. 4, 1995, Journal of Alternative and Complementary Medicine, Vol. 1, pp. 361-369.

13. Randomized Study of the Efficacy and Safety of Oral Elderberry Extract in the Treatment of Influenza A and B Virus Infections . Zakay-Rones, Z, et al. 2, 2004, Journal of International Medical Research, Vol. 32, pp. 132-140.

14. Elderberry Supplementation Reduces Cold Duration and Symptoms in Air-Travellers: A Randomized, Double-Blind Placebo-Controlled Clinical Trial. Tiralongo, Evelin, Wee, Shirley S and Lea, Rodney A. 4, 2016, Nutrients, Vol. 8, p. 182.

Why the Whole Berry Works Better Than a Pill

By Susan Elrod, PhD

As much as you might love the flavor of elderberries and other fruits, many people consider taking supplements to gain the benefits from plant-based polyphenols.

After all, why not? If those polyphenols are so wonderful for us, why not get more of them by taking a concentrated amount in a pill?

The answer is pretty complicated. Individual studies of supplementation of antioxidants and/or extracts may show benefits in some conditions. However, in order for such supplementation to be recommended to the general population, the majority of evidence must support the benefits of supplements.

In some cases, supplementation is highly recommended, as with iron supplementation for children and infants at risk for anemia, or calcium supplementation for pregnant women.(1,2) In terms of vitamin or antioxidant supplementation, however, the evidence simply isn't there. The strongest evidence for plant-based prevention of cardiovascular disease and cancer is in consumption of a wide range of fruits and vegetables, whereas there is not enough evidence to recommend supplementation for this purpose.(3) In fact, experts specifically recommend against supplementing with beta-carotene, Vitamin E, or selenium for prevention of cardiovascular disease or cancer.(3,4)

So why is it that, if polyphenols and antioxidants are so good for us, it's best not to use supplements? Well, as discussed in an earlier article, there are myriad polyphenols in any given fruit or vegetable, and it can be difficult to discern which one is responsible for benefits associated with consumption of that fruit or vegetable.

In fact, it is far more likely that the benefit of a fruit or vegetable is due to the complex interplay among the various polyphenols, rather than one or even a few individual polyphenols. In medicine, a synergistic effect occurs with some compounds: the combination will be more effective than either alone. With such a wide range of polyphenols in plants, it is highly likely that such synergy may yield medical benefits, and that use of a single polyphenol will suppress that benefit.(5)

Incidentally, this is why it is advisable to consume a wide range of fruits and vegetables: the greater diversity of your plant-based diet, the greater diversity of your polyphenol consumption and the greater likelihood that you'll derive benefit from multiple different mechanisms. Furthermore, whole foods contain fiber and other compounds that may contribute to the food's medical benefit, especially in the case of cardiovascular benefits.(6)

So as wonderful as those polyphenols are for you, it's better to get them the more delicious way: in the whole fruit or vegetable rather than in a pill.

Photo Credit: Shutterstock

REFERENCES

1. National Guideline Clearinghouse (NGC). Guideline summary: WHO guideline: daily iron supplementation in infants and children. In: National Guideline Clearinghouse (NCG) [Web site]. Rockville (MD): Agency for Healthcare Research and Quality (AHRQ); 2016 Jan 01. [cited 2017 Oct 06]. Available: https://www.guideline.gov

2. National Guideline Clearinghouse (NGC). Guideline summary: Calcium supplementation in pregnant women. In: National Guideline Clearinghouse (NGC) [Web site]. Rockville (MD): Agency for Healthcare Research and Quality (AHRQ); 2013 Jan 01. [cited 2017 Oct 06]. Available: https://www.guideline.gov

3. National Guideline Clearinghouse (NGC). Guideline summary: Vitamin, mineral, and multivitamin supplements for the primary prevention of cardiovascular disease and cancer: U.S. Preventive Services Task Force recommendation statement. In: National Guideline Clearinghouse (NGC) [Web site]. Rockville (MD): Agency for Healthcare Research and Quality (AHRQ); 2014 Feb 01. [cited 2017 Oct 06]. Available: https://www.guideline.gov

4. National Guideline Clearinghouse (NGC). Guideline summary: Risk reduction of prostate cancer with drugs or nutritional supplements. In: National Guideline Clearinghouse (NGC) [Web site]. Rockville (MD): Agency for Healthcare Research and Quality (AHRQ); 2012 May 17. [cited 2017 Oct 06]. Available: https://www.guideline.gov

5. Williamson, E.M. Synergy and other interactions in phytomedicines. Phytomedicine, 2001, 8(5): 404-409.

6. World Health Organization (WHO). Increasing fruit and vegetable consumption to reduce the risk of noncommunicable diseases. In: World Health Organization (WHO) [Web site]. 15 June 2017. [cited 2017 Oct 06]. Available: http://www.who.int/elena/titles/fruit_vegetables_ncds/en/

Are Elderberry Polyphenols Destroyed by Heat?

By Susan Elrod, PhD

You might have heard the amount of polyphenols change based on whether you're consuming the whole fruit, juice, or some other processed product. In some ways this change is positive; since a lot of fruit goes into juice, there may be a higher phenolic concentration in a serving of juice than in a serving of fruit. Furthermore, heating or other processing techniques makes some polyphenols more bioavailable (that is, usable by the body); lycopene from tomatoes and beta-carotene from carrots are two famous examples of this phenomenon.(1,3)

But what about elderberries and other dark berries? Will those wonderful anthocyanins still be available to you in the form of juice or other processed products? If, like me, you enjoy a mug of mulled wine in the winter, are you destroying the beneficial polyphenols by heating?

Like a lot of topics related to diet and nutrition, the answer to those questions is quite complicated. First, it's important to distinguish between simply heating the fruit and cooking the fruit. Heat itself may destroy certain polyphenols, but the incorporation of other ingredients (particularly yeast when baking) may actually stabilize and thus preserve certain compounds during the baking process.

One study found the process of using heat to dry blueberries resulted in the loss of up to 55% of the initial polyphenols, and up to 95% loss of the initial anthocyanins. (4) However, another study found that baking blueberries with yeast and other leavening agents acted to preserve or even increase polyphenolic compounds. (5) The difference here is that it's not just heat that can affect polyphenols; pH and other factors affected by yeast can also affect the stability of such compounds.

So if a fruit product is simply heated, not incorporated into a baked food, it appears that it may decrease the amount of anthocyanins and related products. However, the impact on total polyphenolic content may not be as significant; a study on Dwarf elderberry found up to 40% reduction in cyanidin-3-glucoside, but only up to 5% decrease in total polyphenols and antioxidants.(6)

Additionally, a study on grape juice found the pasteurization process increased anthocyanins and antioxidants.(7) It appears, then, that while you may not get the absolute best concentration of anthocyanins by heating dark berry products, there are ways to overcome that. Additionally, you may not have to worry quite as much about the total phenolic or antioxidant content decreasing due to heat. The best course of action may be to avoid heating if possible, but don't worry too much about it if you have a warm fruit beverage on occasion.

We mentioned the issue of juicing above. As a larger amount of fruit goes into a single serving of juice, that may be an easy and quick way to get your fruit and/or vegetable servings. Studies on the juicing process in blueberries found that smaller polyphenolic compounds tend to be retained during this process, whereas larger polyphenols tend to be degraded. Blanching the fruit before juicing may also prevent polyphenolic loss.(5) Additionally, several studies conducted on elderberry and other fruits used juice rather than whole fruit, typically because experiments conducted during such studies require the fruit to be in a liquid form.(8,9)

In the case of human studies, juices may be used due to ease of dosing and other experimental controls.(10,11) As such, while the polyphenolic profile may change due to juicing, the total polyphenols and benefits associated with such fruits may not be strongly impacted.

Photo Credit: Shutterstock

REFERENCES

1. Agarwal A, Shen H, Agarwal S, Rao A. Lycopene content of tomato products: its stability, bioavailability and in vivo antioxidant properties. Journal of medicinal food 2001;4(1):9-15.

2. Kamiloglu S, Demirci M, Selen S, Toydemir G, Boyacioglu D, Capanoglu E. Home processing of tomatoes (Solanum lycopersicum): effects on in vitro bioaccessibility of total lycopene, phenolics, flavonoids, and antioxidant capacity. J Sci Food Agric 2014;94(11):2225-2233.

3. Livny O, Reifen R, Levy I, Madar Z, Faulks R, Southon S, et al. Beta-carotene bioavailability from differently processed carrot meals in human ileostomy volunteers. Eur J Clin Nutr 2003;42(6):338-345.

4. Zielinska M, Michalska A. Microwave-assisted drying of blueberry (Vaccinium corymbosum L.) fruits: Drying kinetics, polyphenols, anthocyanins, antioxidant capacity, colour and texture. Food Chemistry 2016 1 December 2016;212(Supplement C):671-680.

5. Rodriguez-Mateos A, Cifuentes-Gomez T, George TW, Spencer JP. Impact of cooking, proving, and baking on the (poly) phenol content of wild blueberry. J Agric Food Chem 2013;62(18):3979-3986.

6. Jimenez P, Cabrero P, Basterrechea JE, Tejero J, Cordoba-Diaz D, Cordoba-Diaz M, et al. Effects of short-term heating on total polyphenols, anthocyanins, antioxidant activity and lectins of different parts of dwarf elder (Sambucus ebulus L.). Plant foods for human nutrition 2014;69(2):168-174.

7. Genova G, Tosetti R, Tonutti P. Berry ripening, pre-processing and thermal treatments affect the phenolic composition and antioxidant capacity of grape (Vitis vinifera L.) juice. J Sci Food Agric 2016;96(2):664-671.

8. Fal AM, Conrad F, Schonknecht K, Sievers H, Pawinska A. Antiviral activity of the "Virus Blocking Factor" (VBF) derived i.a. from Pelargonium extract and Sambucus juice against different human-pathogenic cold viruses in vitro. Wiad Lek 2016;69(3 pt 2):499-511.

9. Effects of Elderberry Juice from Different Genotypes on Oxidative and Inflammatory Responses in Microglial Cells. I International Symposium on Elderberry 1061; 2013.

10. Frank T, Sonntag S, Strass G, Bitsch I, Bitsch R, Netzel M. Urinary pharmacokinetics of cyanidin glycosides in healthy young men following consumption of elderberry juice. Int J Clin Pharmacol Res 2005;25(2):47-56.

11. Murkovic M, Abuja P, Bergmann A, Zirngast A, Adam U, Winklhofer-Roob B, et al. Effects of elderberry juice on fasting and postprandial serum lipids and low-density lipoprotein oxidation in healthy volunteers: a randomized, double-blind, placebo- controlled study. Eur J Clin Nutr 2004;58(2):244-249.

Section 3: How Elderberry Supports Our Health and Healing

An Overview of Elderberry's Health Benefits

By Sarah Crawford, PhD

It all started with Hippocrates, the Ancient Greek "father of medicine," who was the first to report on the wide range of healing effects of the elderberry plant (1). Since then, the elderberry has been used over the centuries to treat a wide variety of ailments, including fevers, wounds, infections and worse! Fast forward to the 21st century: what are scientists saying now? Has the elderberry continued to live up to its long-standing reputation as a healing agent of great and varied dimensions? It can be very difficult to determine the mechanism of action of a specific drug or therapeutic by direct patient assessment.

It is necessary to take the system apart, open the cell and use molecular tools to define the precise target by which a novel agent exerts its therapeutic effects. Ideally, lab tests and patient clinical trials should converge to generate a cause and effect explanation of the therapeutic properties of an investigational agent.

Researchers are currently studying in the laboratory (in vitro) the bioactive components of the elderberry that impact human health. Over the last decade, the research focus has shifted from a more general descriptive assessment of broad-spectrum health benefits to a more molecular-based identification of the specific phytochemicals present in the elderberry that may play a direct cause-and-effect role in human physiology. This paper will attempt to bring together what is currently known about the biological properties of the elderberry subject to laboratory assessment and how these data complement or contradict what has been observed in human patients.

Elderberry v. Flu

Powerful laboratory in vitro (meaning studies in cells that are not in the human body) evidence exists in support of clinical studies in which patients who took elderberry extract preparations at the onset of the flu experienced milder symptoms and shorter duration of the illness.

Viruses cause infections by entering healthy cells and destroying them. Researchers have shown that Sambucol extracts prepared from elderberries block flu viruses from 10 different strains from infecting human cells in culture (2). This research in cell culture in vitro showed conclusively that the elderberry extract binds to the virus and prevents it from attaching to human cells. In one of these in vitro studies, Roshek et al. showed that two flavonoids from elderberry are responsible for this effect, a methylated quercetin and a flavonone derivative (2).

These researchers found that the virus-inactivating properties of the elderberry compounds were similar to the anti-viral medications currently in use, Tamiflu, and Amantadine. Viewed together, the anti-viral properties of these specific elderberry phytochemicals observed in the test tube corroborate patient data showing that elderberry is a beneficial treatment for the flu. When a laboratory research project shows consistent and reproducible molecular results, this is firm evidence of a documented mechanism of action. In this case, the evidence supports the conclusion that the elderberry bioactive components directly block infection by the flu virus.

Photo Credit: Shutterstock

Immune System Boosters: Working Together

One of the most abundant compounds found in elderberries is anthocyanins, plant pigments that give the berry its deep purple hue. Anthocyanins are well known for their anti-oxidant properties; they act by neutralizing the cell damaging effects of free radicals.

In addition, anthocyanins have been found to boost immune system function, by stimulating the production of immune system activators called cytokines (3). These activating cytokines are signals that stimulate the immune system to fight infection. In one study, blood samples were taken from twelve healthy donors and tested for these immune system activators before and after consuming elderberry extracts prepared from Sambucus nigra, the black elderberry.

These studies provided direct evidence that immune system activation is the result of the effects of anthocyanins on the production of cytokines. Moreover, these findings suggest that elderberries may enhance immune system function to provide protection against infections as an immunoprotective agent. Agents that work in this way may have protective health benefits against chronic inflammatory diseases, such as heart disease and cancer.

Stress Protecting Chemicals

In another study to assess the stress-fighting effects of specific elderberry components in the human body, eight healthy human subjects were given a single dose of elderberry juice (4). Plasma levels of anti-oxidants and total phenolics, about ⅔ of which were anthocyanins, increased significantly within one hour after patients consumed 400 mL of juice.

This study showed that the phenolic component of elderberries, especially anthocyanins, are associated with stress-fighting anti-oxidant effects in the body. Simonyi et al. compared the anti-inflammatory effects of an extract prepared from Sambucus nigra canadensis to specific purified components of the elderberry: anthocyanins and the flavonols quercetin and rutin, in mouse brain tissue immune cells (microglial) cells (5).

These studies showed that the anthocyanins found in the elderberry are largely responsible for its anti-oxidant properties. Moreover, the quercetin fraction of the elderberry was found to be primarily responsible for reducing oxidative stress in the microglial cells. Recently, Glensk et al. identified two triterpenoids, ursolic acid, and oleanolic acid, from elderberries (6). The levels of ursolic acid were found to be threefold that of oleanolic acid. Further studies suggest that these compounds may block the growth of tumor cells (7).

Conclusion

An English scientist in the 17th century, John Evelyn, once proclaimed, "If the medicinal properties of the elderberry were fully known, I cannot tell what our countrymen could ail from which he might not fetch a remedy, either for sickness or wounds" (8).

Numerous in vitro studies conducted over the past decade have provided convincing evidence of the anti-inflammatory, free radical scavenging and immune system promoting activities of several bioactive compounds from the elderberry, including anthocyanins, quercetin, and other phenolic compounds. Moreover, both laboratory studies and patient clinical trial data confirm a direct cause-and-effect explanation for the anti-viral properties of this fruit.

What we need now are additional studies attempting to further pinpoint the identity and specific biological effects of the elderberry phytochemicals. Each of these bioactive components must be tested individually and in combination in the lab for specific targeted dose/responses.

The functions of the immune system and oxidative stress responses are tightly intertwined and are at the crossroads of many complex human diseases. As we learn more and more about the broad spectrum physiological effects of the elderberry's bioactive components, maybe we will find that the ancients were right after all.

Photo Credit: Shutterstock

Table 10: Summary of Biologically Active Components of the Elderberry

ELDERBERRY PHYTOCHEMICALS CATEGORY	BIOACTIVE COMPOUND	BIOLOGICAL TARGET
Flavonoids	Methyl quercetin	Binds and inactivates flu virus
Flavonoids	Flavonone derivative	Binds and inactivates flu virus
Flavonoids	Anthocyanins	Immune System activation
Flavonoids	Anthocyanins	Anti-oxidant activity
Flavonoids	Quercetin	Reduces oxidative stress in brain microglial cells
Triterpenoids	Ursolic acid	Reduces oxidative stress, blocks cell growth
Triterpenoids	Oleanolic acid	Reduces oxidative stress, blocks cell growth

REFERENCES

1. Totelin, Laurence MV, ed. Hippocratic recipes: oral and written transmission of pharmacological knowledge in fifth-and fourth-century Greece. Vol. 34. Brill, 2009.

2. Roschek, Bill, et al. "Elderberry flavonoids bind to and prevent H1N1 infection in vitro." Phytochemistry 70.10 (2009): 1255-1261.

3. Barak, Vivian, Tal Halperin, and Inna Kalickman. "The effect of Sambucol, a black elderberry-based, natural product, on the production of human cytokines: I. Inflammatory cytokines." Eur Cytokine Netw 12.2 (2001): 290-296.

4. Netzel, Michael, et al. "In vivo antioxidative capacity of a composite berry juice." Food research international 35.2 (2002): 213-216.

5. Simonyi, Agnes, et al. "Inhibition of microglial activation by elderberry extracts and its phenolic components." Life sciences 128 (2015): 30-38.

6. Gleńsk, Michał, et al. "Determination of ursolic and oleanolic acid in Sambuci fructus." Chemistry & biodiversity 11.12 (2014): 1939-1944.

7. Gleńsk, Michał, et al. "Triterpenoid Acids as Important Antiproliferative Constituents of European Elderberry Fruits." Nutrition and Cancer 69.4 (2017): 643- 651.

8. Leong, Elaine. "Making medicines in the early modern household." Bulletin of the History of Medicine 82.1 (2008): 145-168.

A Comprehensive Review of Elderberry's Effects, Scientific Studies Using Elderberry, 2013 – 2018

By Nnenna Udensi, MD

This review covers the recent literature regarding the effectiveness and safety of elderberry products in medicine. Its purpose is to discuss and update the findings in previous literature reviews like Ulbricht(1) and Porter(18). The majority of the studies referenced will be from 2013-2018, although, older studies may be included as well for completeness. Publications not referenced directly in this article will still be incorporated in appendix charts and will contain the disease, lab results, and animal studies not addressed in this review. Including this information may hopefully serve to further the conversation surrounding elderberry use in medicine.

My review of elderberry studies since 2013, conducted using various databases (including Pubmed.org, herbmed.org, herbalgram.org) yielded 13 new studies on safety and effectiveness in humans, ten recent animal studies, and over 60 studies involving laboratory, chemical, and genetic studies of elderberry.

This article will only address those studies related to our inquiry, a fraction of the available data, but the complete list of studies is available in Appendix A. I refer to some studies older than 2013 in the article, but the charts will only include publications from 2013-2018 since earlier studies are referenced in other reviews(1,18). I did not have access to the full texts of all the publications, so in those cases, I cited the abstract alone.

This article will focus on ten common ailments that the previous reviews(1, 18) and studies suggest elderberry may affect. Although there are additional potential benefits that have been or are currently under investigation, the ones listed here are more likely to apply to the daily lives of average people. Elderberry has several useful medical applications, and the goal of this article is to present those medical uses and the evidence supporting them to you the reader.

What is Elderberry

Sambucus nigra, more commonly known as elderberry, is a flowering plant found in subtropical regions worldwide. It produces large clusters of dark-colored berries. People consume the fruit and its products for a variety of reasons including dietary supplementation, remedies for various ailments, and regular consumption as food. Ripe, cooked elderberries are edible; however, raw fruit and other parts of the plant can be toxic in sufficient quantities(23).

While similar to other berries, in everyday use elderberry is unique in that it has a wealth of potential health and medical benefits and applications not often associated with different types of berries. Some of the benefits ascribed to elderberries include:

- Reduction in cold symptoms
- Constipation
- Bacterial sinusitis/ bronchitis Allergies
- Diuretic effect for urination
- Gingivitis Hyperlipidemia Obesity
- Anti-inflammatory and anti-microbial effects Overall wellbeing

The purported health benefits of elderberry have varying degrees of anecdotal, scientific, and clinical trial support. The purpose of my research effort was to review all scientific studies published between January 2013 and July 2018 that sought to investigate different medical aspects of elderberry.

After reviewing each study, I found that some of the evidence shows clear health benefits, while other data is mixed. So, this article will discuss the merits of these and other claims about the effectiveness of elderberry as medicine. For simplicity's sake, a rating system cited by Ulbricht(1) will be used to describe elder's effect for each disease.

Reduction In Cold Symptoms

Flu symptoms relief is one of the more well-studied benefits of elderberry, and there is a lot of useful data to support it. The antiviral properties are well documented. One study(2) found that elderberry extract reduced the duration of symptoms (aches, pain, cough, congestion, sleep and mucus discharge) by about half, 3-4 days down from 7-8 days. A similar study(3) observed the effects of elderberry in the first 48 hours of confirmed influenza infection. It also showed marked improvement in shorter periods of time.

These and other studies(4) suggest that if taken early in the infection, elderberry extract can safely be used to reduce the time and severity of flu symptoms. Since 2013 four studies, including one clinical trial and three laboratory studies, were conducted and the results confirm these and Ulbricht(1) findings.

Bacterial Sinusitis/Bronchitis

Sinusitis is characterized by inflammation of the tissue that lines the sinus cavity. Likewise, bronchitis is characterized by inflammation of the bronchial lining in the lungs. The causes of the inflammation can include viruses, bacteria, and allergens among other things.

The symptoms often include pain, fever, drainage, cough, and fatigue. When tested(8) against bacteria that commonly cause these conditions, elderberry was shown to have antimicrobial properties. There were six studies during the five and half year review period and the results of these studies support the findings from previous reviews that show elderberries' effectiveness against respiratory bacteria and viruses.

Photo Credit: Shutterstock

Allergies

Allergic rhinitis is characterized by nasal congestion or drainage, sneezing, and itching of the nose, ears, or eyes. The causes include various allergens such as pollens, animal dander, dust, etc.

There's very little new scientific support for this claim during the 2013 – 2018 period. While studies(11) show that elder appears in alternative treatments for allergic rhinitis, there are currently no studies that focus on the treatment of allergies with elderberry. Since the 2013 observational study falls within the review period, I felt it appropriate to address the claim.

Gingivitis

Gingivitis is mild gum disease characterized by irritation, redness, and swelling of the gum around the base of teeth. Left untreated, it can lead to more severe disease.

Studies(13,14) examining the anti-inflammatory effects of elder reveal useful anti-inflammatory properties. One paper suggests that these properties of elderberry are helpful for conditions like gingivitis that have inflammation as their cause and that elderberry is excellent for overall oral health. In fact, mouthwash containing an elderberry preparation is currently in clinical trials(22).

Hyperlipidemia

Hyperlipidemia refers to excess fat in the blood. There is some evidence(15) from animal studies conducted during the review period that elderberry extract can reverse hyperlipidemia, but the research is limited in scope, and there is no evidence so far of this effect in humans. One study(16) done on human cells saw no effect at all.

Obesity

Obesity, generally regarded as excessive energy storage caused by increased calorie intake, has inspired new questions about its causes and effects. Research is starting to take a harder look at how things like inflammation affect obesity and chronic diseases associated with it.

Elderberry extract and similar fruits have been shown to be effective at reducing inflammation in several studies(20) conducted from 2013 to 2018. So, given the links between obesity and inflammation; and inflammation and elderberry, one study sought to examine potential connections between elderberry and obesity directly. The conclusions(17) drawn by the scientists were limited, but they noted the fact that elderberry seemed to attenuate obesity driven inflammation. So even if elderberry doesn't directly cause weight reduction, if it reduces potential side effects of obesity it could be invaluable in a health maintenance capacity.

Inflammation

Inflammation is one of the body's defense mechanisms. The inflammatory process combats irritants and pathogens while allowing the body to begin healing. However, when inflammation is chronic, it can go from being part of the solution to being the problem. Conditions like COPD, arthritis, Crohn's disease, and obesity all have inflammation at the core of the disease process.

While there is no evidence supporting elderberry treatment in some of these more serious diseases, there are seven studies in the last five years that look at several less complicated conditions, some discussed in this paper, that have inflammatory disease processes that may benefit from elderberry therapies and its reductive effect on inflammation(17). The overall consensus of these studies is that elderberry can reduce inflammation. Future studies will likely address questions about whether this effect is enough to change the course of chronic inflammatory diseases. As of now, it is still unclear.

Constipation

Loosely defined, constipation is infrequent or difficult bowel movements. Causes include other disease, medications, or various factors such as lifestyle or diet.

Fortunately, in addition to the range of conventional medical therapies to treat constipation, there is evidence(7) that elderberry is useful as a laxative. When served as a tea, a study showed a significant decrease in constipation symptoms.

As is common with other laxatives, one of the side effects noted in the study is cramping. However, when combined with other herbs such as fennel, the study suggested that cramping caused by elderberry can be reduced or eliminated(7).

Diuresis

Diuresis is defined as increased or excessive production of urine. Factors include things as benign as drinking too much water, to more severe conditions such as chronic disease. Substances or medications that cause this are called diuretics.

A healthy person would not typically need to induce diuresis, but people with conditions like hypertension may benefit from diuresis. There are older more traditional forms of medicine that use elder in this way. However, there is very little to no scientific evidence to back up this claim, and there appear to be no recent efforts to study this effect.

Overall Wellbeing

This review is not, by any means an exhaustive list of the health benefits of elderberry; more so an overview of the more tangible effects that have at least some scientific backing. Traditional medicine and historical anecdotes extol the virtues of elder in a variety of home remedies for a multitude of ailments. Modern medicine, while seemingly slow to catch up, is increasingly finding elder useful as an alternative or complementary medicine. It's shown to be safe for consumption in a variety of different ways. So long as one does not eat the plant raw, ingesting elderberry is not toxic. One caveat, people who have allergies to plants similar to elder such as honeysuckle, may have allergic reactions to elderberry products.

In a study of participants seeking a healthy lifestyle and weight changes, the addition of elderberry to their new regimen significantly improved their emotional well being and quality of life(20,21) In addition, well documented anti-oxidant properties can make elderberry useful overall as part of a healthy diet and lifestyle(20,21).

Recent Studies & Clinical Trials

Other studies(25-33) conducted during the review period 2013-2018 have shown positive effects involving cholesterol and LDL, atherosclerosis, wound care, anti-cancer activities and more! These were not covered in this review because they either have conflicting or inconclusive data, or they have not been replicated or peer-reviewed. The benefit of corroborating data for claims like these allows for more intelligent questions and decisions when it comes to science and healthcare.

These claims are too recent to have a lot of corroborating data for them yet, but they are intriguing findings and are likely to be studied in the future. So although more studies are needed, the possibilities are exciting! There are currently three active studies, registered in the clinicaltrials.gov database, involving elderberry and influenza, a mouthwash, and human immunity.

Endnotes

I included two conditions – constipation and diuresis – with studies older than 2013;. I found the documented laxative effects for constipation to be more robust than the view espoused in the Ulbricht paper. Diuresis was not addressed by Ulbricht so I addressed it here. In this article, I cited but did not refer to the views expressed in the Ulbricht paper specifically. I chose to present the data for the 2013 – 2018 period as a whole. Only a few of my conclusions contrasted with those of the Ulbricht review. I noted these few differences in the chart that makes up most of Appendix A.

I did not see where Ulbricht made any claims either way about allergies and treatment with elderberry except to mention potential allergic reactions to elderberry itself. I didn't see anything specific in the citations either. Independently, the only study referencing allergies that I found was the 2013 study that I cited in this article. If you have other scientific data about treating this condition with elderberry, please email me at nudensi@gmail.com so I can add it to the review.

REFERENCES

1. Ulbricht C, Basch E, Cheung L, et al. An Evidence-Based Systematic Review of Elderberry and Elderflower (Sambucus nigra) by the Natural Standard Research Collaboration. Journal of Dietary Supplements. 2014;11(1):80-120. doi:10.3109/19390211.2013.859852.

2. Zakay-Rones Z, Thom E, Wollan T, Wadstein J. Randomized Study of the Efficacy and Safety of Oral Elderberry Extract in the Treatment of Influenza A and B Virus Infections. Journal of International Medical Research. 2004;32(2):132-140. doi:10.1177/147323000403200205.

3. Zakay-Rones Z, Varsano N, Zlotnik M, et al. Inhibition of Several Strains of Influenza Virus in Vitro and Reduction of Symptoms by an Elderberry Extract (Sambucus nigra L.) during an Outbreak of Influenza B Panama. The Journal of Alternative and Complementary Medicine. 1995;1(4):361-369. doi:10.1089/acm.1995.1.361.

4. Guo R, Pittler MH, Ernst E. Complementary Medicine for Treating or Preventing Influenza or Influenza-like Illness. The American Journal of Medicine. 2007;120(11). doi:10.1016/j.amjmed.2007.06.031.

5. Roxas M, Jurenka J. Colds and influenza: a review of diagnosis and conventional, botanical, and nutritional considerations. Alternative Medicine Review. 2007;12(1):25- 48.

6. Abuja PM, Murkovic M, Pfannhauser W. Antioxidant and Prooxidant Activities of Elderberry (Sambucus nigra) Extract in Low-Density Lipoprotein Oxidation. Journal of Agricultural and Food Chemistry. 1998;46(10):4091-4096. doi:10.1021/jf980296g.

7. Picon PD, Picon RV, Costa AF, et al. Randomized clinical trial of a phytotherapic compound containing Pimpinella anisum, Foeniculum vulgare, Sambucus nigra, and Cassia augustifolia for chronic constipation. BMC Complementary and Alternative Medicine. 2010;10(1). doi:10.1186/1472-6882-10-17.

8. Krawitz C, Mraheil MA, Stein M, et al. Inhibitory activity of a standardized elderberry liquid extract against clinically-relevant human respiratory bacterial pathogens and influenza A and B viruses. BMC Complementary and Alternative Medicine. 2011;11(1). doi:10.1186/1472-6882-11-16.

9. Knudsen B, Kaack K. A Review Of Traditional Herbal Medicinal Products With Disease Claims For Elder (Sambucus Nigra) Flower. Acta Horticulturae. 2015;(1061):109-120. doi:10.17660/actahortic.2015.1061.11.

10. Hearst C. Antibacterial activity of elder (Sambucus nigra L.) flower or berry against hospital pathogens. Journal of Medicinal Plants Research. 2010;4(17):1805-1809. doi:10.5897/JMPR10.147.

11. Sayin I, Cingi C, Oghan F, Baykal B, Ulusoy S. Complementary Therapies in Allergic Rhinitis. ISRN Allergy. 2013;2013:1-9. doi:10.1155/2013/938751.

12. Wright C, Van-Buren L, Kroner C, Koning M. Herbal medicines as diuretics: A review of the scientific evidence. Journal of Ethnopharmacology. 2007;114(1):1-31. doi:10.1016/j.jep.2007.07.023.

13. Levine WZ, Samuels N, Sheshet MEB, Grbic JT. A Novel Treatment of Gingival Recession using a Botanical Topical Gingival Patch and Mouthrinse. The Journal of Contemporary Dental Practice. 2013;14:948-953. doi:10.5005/jp-journals-10024- 1431.

14. Harokopakis E, Albzreh MH, Haase EM, Scannapieco FA, Hajishengallis G. Inhibition of Proinflammatory Activities of Major Periodontal Pathogens by Aqueous Extracts From Elder Flower (Sambucus nigra). Journal of Periodontology. 2006;77(2):271-279. doi:10.1902/jop.2006.050232.

15. Dubey P, Jayasooriya AP, Cheema SK. Fish oil induced hyperlipidemia and oxidative stress in BioF1B hamsters is attenuated by elderberry extract. Applied Physiology, Nutrition, and Metabolism. 2012;37(3):472-479. doi:10.1139/h2012-030.

16. Vauzour D, Tejera N, Oneill C, et al. Anthocyanins do not influence long-chain n-3 fatty acid status: studies in cells, rodents and humans. The Journal of Nutritional Biochemistry. 2015;26(3):211-218. doi:10.1016/j.jnutbio.2014.09.005.

17. Lee Y-M, Yoon Y, Yoon H, Park H-M, Song S, Yeum K-J. Dietary Anthocyanins against Obesity and Inflammation. Nutrients. 2017;9(10):1089. doi:10.3390/nu9101089.

18. Porter RS, Bode RF. A Review of the Antiviral Properties of Black Elder (Sambucus nigraL.) Products. Phytotherapy Research. 2017;31(4):533-554. doi:10.1002/ptr.5782.

19. Akram M, Tahir IM, Shah SMA, et al. Antiviral potential of medicinal plants against HIV, HSV, influenza, hepatitis, and coxsackievirus: A systematic review. Phytotherapy Research. 2018;32(5):811-822. doi:10.1002/ptr.6024.

20. Chrubasik C, Maier T, Dawid C, et al. An observational study and quantification of the actives in a supplement withSambucus nigraandAsparagus officinalisused

for weight reduction. Phytotherapy Research. 2008;22(7):913-918. doi:10.1002/ptr.2415.

21. Shahsavandi S, Ebrahimi MM, Hasaninejad Farahani A. Interfering With Lipid Raft Association: A Mechanism to Control Influenza Virus Infection By Sambucus Nigra. Iranian Journal of Pharmaceutical Research : IJPR. 2017;16(3):1147-1154.

22. Evaluation and Comparison of Efficacy of PeriActive Mouthwash to Chlorhexidine 0.12% Mouth Rinse. Full Text View - ClinicalTrials.gov. https://clinicaltrials.gov/ct2/show/NCT02987634?term=Sambucus nigra&rank=3. Accessed August 6, 2018.

23. Tejero J, Jiménez P, Quinto E, et al. Elderberries: A Source of Ribosome-Inactivating Proteins with Lectin Activity. Molecules. 2015;20(2):2364-2387. doi:10.3390/molecules20022364

24. Bear S, Mohanasundaram J, Axentiev P. HerbMed. http://www.herbmed.org/. Accessed August 20, 2018.

25. Kaptan E, Sancar-Bas S, Sancakli A, Bektas S, Bolkent S. The effect of plant lectins on the survival and malignant behaviors of thyroid cancer cells. J Cell Biochem. 2018 Jul;119(7):6274-6287. doi: 10.1002/jcb.26875. Epub 2018 Apr 16. PubMed PMID: 29663501

26. Jarić S, Kostić O, Mataruga Z, Pavlović D, Pavlović M, Mitrović M, Pavlović P. Traditional wound-healing plants used in the Balkan region (Southeast Europe). J Ethnopharmacol. 2018 Jan 30;211:311-328. doi: 10.1016/j.jep.2017.09.018. Epub 2017 Sep 21. Review. PubMed PMID: 28942136.

27. Nilsson A, Salo I, Plaza M, Björck I. Effects of a mixed berry beverage on cognitive functions and cardiometabolic risk markers; A randomized cross-over study in healthy older adults. PLoS One. 2017 Nov 15;12(11):e0188173. doi: 10.1371/journal.pone.0188173. eCollection 2017. PubMed PMID: 29141041; PubMed Central PMCID: PMC5687726.

28. Gleńsk M, Czapińska E, Woźniak M, Ceremuga I, Włodarczyk M, Terlecki G, Ziółkowski P, Seweryn E. Triterpenoid Acids as Important Antiproliferative Constituents of European Elderberry Fruits. Nutr Cancer. 2017 May-Jun;69(4):643-651. doi: 10.1080/01635581.2017.1295085. Epub 2017 Mar 21. PubMed PMID: 28323490.

29. Salvador ÂC, Król E, Lemos VC, Santos SA, Bento FP, Costa CP, Almeida A, Szczepankiewicz D, Kulczyński B, Krejpcio Z, Silvestre AJ, Rocha SM. Effect of Elderberry (Sambucus nigra L.) Extract Supplementation in STZ-Induced Diabetic Rats Fed with a High-Fat Diet. Int J Mol Sci. 2016 Dec 22;18(1). pii: E13. doi:

10.3390/ijms18010013. PubMed PMID: 28025494; PubMed Central PMCID: PMC5297648.

30. Kirichenko TV, Sobenin IA, Nikolic D, Rizzo M, Orekhov AN. Anti-cytokine therapy for prevention of atherosclerosis. Phytomedicine. 2016 Oct 15;23(11):1198-210. doi: 10.1016/j.phymed.2015.12.002. Epub 2015 Dec 19. Review. PubMed PMID: 26781385.

31. Schröder L, Richter DU, Piechulla B, Chrobak M, Kuhn C, Schulze S, Abarzua S, Jeschke U, Weissenbacher T. Effects of Phytoestrogen Extracts Isolated from Elder Flower on Hormone Production and Receptor Expression of Trophoblast Tumor Cells JEG-3 and BeWo, as well as MCF7 Breast Cancer Cells.

32. Farrell N, Norris G, Lee SG, Chun OK, Blesso CN. Anthocyanin-rich black elderberry extract improves markers of HDL function and reduces aortic cholesterol in hyperlipidemic mice. Food Funct. 2015 Apr;6(4):1278-87. doi: 10.1039/c4fo01036a. PubMed PMID: 25758596.

33. El-Houri RB, Kotowska D, Olsen LC, Bhattacharya S, Christensen LP, Grevsen K, Oksbjerg N, Færgeman N, Kristiansen K, Christensen KB. Screening for bioactive metabolites in plant extracts modulating glucose uptake and fat accumulation. Evid Based Complement Alternat Med. 2014;2014:156398. doi: 10.1155/2014/156398. Epub 2014 Aug 28. PubMed PMID: 25254050; PubMed Central PMCID: PMC4164421.

APPENDIX A

Description

This appendix should provide a comprehensive listing of all scientific studies involving elderberry that were conducted and reported in Pubmed.org, herbmed.org, and herbalgram.org from January 1, 2013 to July 1, 2018.

My review of elderberry studies since 2013 yielded 13 new studies on safety and effectiveness in humans, ten recent animal studies, and over 60 studies involving laboratory, chemical, and genetic studies of elderberry.

For completeness, I have included a number of studies in this chart that I did not directly reference in the full body of the report. Including this information may hopefully serve to further the conversation surrounding elderberry use in medicine.

Scoring System (Level of Evidence Grade 1)

To convey my assessment of how each study affects the overall level of scientific consensus for each health topic, I used the scoring system applied in Ulbricht.

A. Strong scientific evidence
B. Good scientific evidence
C. Unclear or conflicting scientificevidence
D. Fair negative scientific evidence
E. Strong negative evidence
F. Lack of evidence

Criteria

A. Statistically significant evidence of benefit from >2 properly randomized trials (RCTs), OR evidence from one properly conducted RCT AND one properly conducted meta-analysis, OR evidence from multiple RCTs with a clear majority of the properly conducted trials showing statistically significant evidence of benefit AND with supporting evidence in basic science, animal studies, or theory.

B. Statistically significant evidence of benefit from 1 to 2 properly randomized trials, OR evidence of benefit from >1 properly conducted meta-analysis OR evidence of benefit from >1 cohort/case-control/nonrandomized trials AND with supporting evidence in basic science, animal studies, or theory.

C. Evidence of benefit from >1 small RCT(s) without adequate size, power, statistical significance, or quality of design by objective criteria,* OR conflicting evidence from multiple RCTs without a clear majority of the properly conducted trials showing

D. evidence of benefit or ineffectiveness, OR evidence of benefit from >1 cohort/case– control/nonrandomized trials AND without supporting evidence in basic science, animal studies, or theory, OR evidence of efficacy only from basic science, animal studies, or theory.

E. Statistically significant negative evidence (i.e., lack of evidence of benefit) from cohort/case–control/nonrandomized trials, AND evidence in basic science, animal studies, or theory suggesting a lack of benefit.

F. Statistically significant negative evidence (i.e., lack of evidence of benefit) from >1 properly randomized adequately powered trial(s) of high-quality design by objective criteria.

G. Unable to evaluate efficacy due to lack of adequate available human data.

HUMAN AND ANIMAL STUDIES 2013 - 2018

Condition	Score	Publication Year	Publication Name	Authors	Findings	Mechanisms/Mode of Action	Elderberry Type	Elder Quantity	Study Type
Thyroid Cancer	C	2018	The effect of plant lectins on the survival and malignant behaviors of thyroid cancer cells.	Kaptan	Strong candidate for developing therapeutic strategies.	Altered cell surface glycosylation	Sambucus nigra agglutinin (SNA)	-	Lab
Anti- viral	B	2018	Antiviral potential of medicinal plants against HIV, HSV, influenza, hepatitis, and coxsackievirus: A systematic review	Akram	Antiviral properties	Promising specific antiviral activities	Sambucus Nigra extract	-	Human
LDL/ Memory	C	2017	Effects of a mixed berry beverage on cognitive functions and cardiometabolic risk markers; A randomized cross-over study in healthy older adults.	Nilsson	Reduced LDL/ Improved Working Memory	-	Berry beverage	50 grams	Human
Obesity	C	2017	Dietary Anthocyanins against Obesity and Inflammation	Lee	Potential regulator of obesity-derived inflammation	Anti-inflammatory effects of anthocyanins	Whole Berry	-	Human
Anti-viral	B	2017	A Review of the Antiviral Properties of Black Elder (Sambucus nigra L.)	Porter	Demonstrated Safety and Antimicrobial Properties	-	Sambucus Nigra extract	-	Human
Wound care	C	2017	Traditional wound-healing plants used in the Balkan region (Southeast Europe).	Jaric	Beneficial effects	-	Infusions, decoctions, tinctures, syrups, oils, ointments, and balms	-	Human

APPENDIX A, CONT.

Condition	Score	Publication Year	Publication Name	Authors	Findings	Mechanisms/Mode of Action	Elderberry Type	Elder Quantity	Study Type
Colds	C	2016	Elderberry Supplementation Reduces Cold Duration and Symptoms in Air-Travellers: A Randomized, Double-Blind Placebo-Controlled Clinical Trial.	Tiralongo	Reduction of cold episodes	-	Sambucus Nigra extract	600-900mg	Human
Seizures	C	2016	Anticonvulsant activities of Sambucus nigra.	Ataee	Anticonvulsant Effects	-	-	250-1000 mcg/kg	Animal
Heavy Metal Toxicity	C	2016	Possible protective role of elderberry fruit lyophilizate against selected effects of cadmium and lead intoxication in Wistar rats.	Kopec	Protective	Increased activity of protective enzymes	Freeze-dried Elderberry Fruits	-	Animal
Insulin Resistance	C	2016	Effect of Elderberry (Sambucus nigra L.) Extract Supplementation in STZ-Induced Diabetic Rats Fed with a High-Fat Diet.	Salvador	Correction of Hyperglycemia	Modulation of glucose metabolism	Sambucus Nigra extract	190-350 mg/kg	Animal
Escherichia Coli	C	2016	Antibacterial activity of fractions from three Chumash medicinal plant extracts and in vitro inhibition of the enzyme enoyl reductase by the flavonoid jaceosidin	Allison	Antibacterial	Enzyme inhibition	Elderberry enzyme Isolates	-	Lab
Hypertension	C	2016	The beneficial effects on blood pressure, dyslipidemia and oxidative stress of Sambucus nigra extract associated with renin inhibitors.	Ciocoiu	Reduced Blood Pressure	Renin–angiotensin system	Sambucus Nigra extract	0.046 g/kg	Animal
Breast Cancer	C	2016	Effects of Phytoestrogen Extracts Isolated from Elder Flower on Hormone Production and Receptor Expression of Trophoblast Tumor Cells JEG-3 and BeWo, as well as MCF7 Breast Cancer Cells.	Schröder	Downregulation of hormone receptors	Modulation of receptor expression	Sambucus Nigra extract	-	Lab
Neuro-degenerative diseases	C	2016	Neuroprotective Role of Natural Polyphenols.	Spagnuolo	Neuroprotective	Antioxidant effects	Sambucus Nigra extract	-	Lab

APPENDIX A, CONT.

Condition	Score	Publication Year	Publication Name	Authors	Findings	Mechanisms/Mode of Action	Elderberry Type	Elder Quantity	Study Type
Atherosclerosis	C	2015	Anti-cytokine therapy for prevention of atherosclerosis	Kichenko	Moderation of disease	Anti-inflammatory	-	-	Human
Fatty Acids	D	2015	Anthocyanins do not influence long-chain n-3 fatty acid status: studies in cells, rodents and humans.	Vauzour	No effect	-	Sambucus Nigra extract	-	Human
Hyperlipidemia	C	2015	Anthocyanin-rich black elderberry extract improves markers of HDL function and reduces aortic cholesterol in hyperlipidemic mice.	Farrell	Cholesterol reduction	Anti-inflammatory	Sambucus Nigra extract	-	Animal
Depression	C	2014	Antidepressant activities of Sambucus ebulus and Sambucus nigra.	Mahmoudi	Antidepressant	-	Sambucus Nigra extract	-	Animal
Inflammatory Bowel Disease	C	2014	Anti-inflammatory activity of fruit fractions in vitro, mediated through toll-like receptor 4 and 2 in the context of inflammatory bowel disease.	Nasef	Anti-inflammatory	Anti-inflammatory	Fractionated Elderberry	-	Lab
Gingivitis	C	2013	A novel treatment of gingival recession using a botanical topical gingival patch and mouthrinse.	Levine	Effective	-	Botanical patch and Rinse	-	Human
Allergic Rhinitis	C	2013	Complementary Therapies in Allergic Rhinitis	Sayin	-	-	Herbal Supplement	-	Case Study

APPENDIX A, CONT.

ADDITIONAL STUDIES ON KINETICS, DYNAMICS, CHEMISTRY, AND GENETICS OF ELDERBERRY. 2013-2018

Condition	Score	Publication Year	Publication Name	Authors	Findings	Elderberry Type	Study Type
Effects of Heat on elderberry proteins	C	2018	The kinetics of thermal degradation of polyphenolic compounds from elderberry	Oancea	Decrease in activity 100 to 150 °C	Extract	Lab
Antimicrobial activity	C	2017	Characterization of Antimicrobial Properties of Extracts of Selected Medicinal Plants	Cioch	Demonstrated antifungal activity	Extract	Lab
Parkinson's Disease	C	2017	Lumbee traditional medicine: Neuroprotective activities of medicinal plants used to treat Parkinson's disease-related symptoms.	De Rus Jacquet	Symptomatic relief, No disease modification	Extract	Human
Anti-Cancer	C	2017	Triterpenoid Acids as Important Antiproliferative Constituents of European Elderberry Fruits.	Glensk	Antiproliferative activities	Extract	Lab
Diabetes	C	2017	Effect of Phenolic Compounds from Elderflowers on Glucose- and Fatty Acid Uptake in Human Myotubes and HepG2-Cells.	Ho	Stimulation of Glucose uptake	Extract	Lab
Inflammatory Disease	C	2017	Elderberry and Elderflower Extracts, Phenolic Compounds, and Metabolites and Their Effect on Complement, RAW 264.7 Macrophages and Dendritic Cells.	Ho	Anti Inflammatory	Extract	Lab
Diabetes	C	2017	Phenolic Elderberry Extracts, Anthocyanins, Procyanidins, and Metabolites Influence Glucose and Fatty Acid Uptake in Human Skeletal Muscle Cells.	Ho	Enhanced glucose Uptake	Extract	Lab
Diabetes	C	2017	Enhanced Glucose Uptake in Human Liver Cells and Inhibition of Carbohydrate Hydrolyzing Enzymes by Nordic Berry Extracts.	Ho	Increased glucose Uptake	Extract	Lab
Urinary Tract Infections	C	2017	MICROTOX TEST AS A TOOL TO ASSESS ANTIMICROBIAL PROPERTIES OF HERBAL INFUSIONS USED IN URINARY TRACT INFECTIONS.	Okragla	Effective	Extract	Human

APPENDIX A, CONT.

Condition	Score	Publication Year	Publication Name	Authors	Findings	Elderberry Type	Study Type
Anti-viral effect	C	2017	Interfering With Lipid Raft Association: A Mechanism to Control Influenza Virus Infection By Sambucus Nigra.	Shahsavandi	Effective at Higher doses	Extract	Lab
Active compound concentration	C	2017	Quantitation of anthocyanins in elderberry fruit extracts and nutraceutical formulations with paper spray ionization mass spectrometry.	Codi	500 to 2370 mg/100 g measured in the dried stems and fruit, respectively.	Extract	Lab
Differentiation between different subspecies of Elderberry	C	2017	Discriminant Analyses of the Polyphenol Content of American Elderberry Juice from Multiple Environments Provide Genotype Fingerprint.	Johnson	Correct identification of 45/48 samples	Extract	Lab
Compound isolation	C	2017	Chlorogenic acid isomer contents in 100 plants commercialized in Brazil.	Meinhart	Chlorogenic acid isolate found in Elderberry	Extract	Lab
Compound isolation	C	2017	Profiling and Quantification of Regioisomeric Caffeoyl Glucoses in Berry Fruits.	Patras	Caffeoyl isolate found in Elderberry	Extract	Lab
Elderberry extract Stability	C	2017	Polyphenolic content, antiradical activity, stability and microbiological quality of elderberry (Sambucus nigra L.) extracts.	Pilszka	Potential as high quality raw material	Extract	Lab
Anti-Cancer	C	2017	Antiproliferative and Apoptotic Potential of Cyanidin-Based Anthocyanins on Melanoma Cells.	Rugina	Melanoma cell death	Extract	Lab
Various preservation Methods	C	2017	Unveiling elderflowers (Sambucus nigra L.) volatile terpenic and norisoprenoids profile: Effects of different postharvest conditions.	Salvador	Vacuum packing and freezing most effective	Extract	Lab
Evaluation of Heavy Metals in raw preparations	C	2017	Transition rates of selected metals determined in various types of teas (Camellia sinensis L. Kuntze) and herbal/fruit infusions.	Schulzki	Varied low levels	Infusion/ Tea	Lab
Antioxidant activity	C	2017	The Phenolic Contents and Antioxidant Activities of Infusions of Sambucus nigra L.	Viapiana	Flower teas more potent then berry teas	Infusion/ Tea	Lab

Condition	Score	Publication Year	Publication Name	Authors	Findings	Elderberry Type	Study Type
Heat Stability	C	2017	Lectin Digestibility and Stability of Elderberry Antioxidants to Heat Treatment In Vitro	Jimenez	Allergic potential reduced with brief heat treatment	Extract	Lab
Immune system stimulation	C	2016	RG-I regions from elderflower pectins substituted on GalA are strong immunomodulators.	Ho	Strong immune modulator	Extract	Lab
Differences in taste compounds	C	2016	Comparison of major taste compounds and antioxidative properties of fruits and flowers of different Sambucus species and interspecific hybrids.	Petkovsek	Significant variation between species and hybrids	Berries	Lab
Differentiation between different subspecies of Elderberry	C	2016	Variation of Select Flavonols and Chlorogenic Acid Content of Elderberry Collected Throughout the Eastern United States.	Mudge	Differentiate American elderberries from wild type	Berries	Lab
Comparison of antioxidant properties of various juices	C	2016	Multidimensional comparative analysis of phenolic compounds in organic juices with high antioxidant capacity.	Nowak	Chokeberry juices had highest concentration of antioxidant properties	Juiced fruit	Lab
Influence of ripening	C	2016	Metabolomic-Based Strategy for Fingerprinting of Sambucus nigra L. Berry Volatile Terpenoids and Norisoprenoids: Influence of Ripening and Cultivar.	Salvador	Higher metabolites in unripe fruit. Decreased as fruit ripens	Pre-harvest fruit	Lab
Bioactive content	C	2016	Effect of variety on content of bioactive phenolic compounds in common elder (Sambucus nigra L.).	Vrchotova	Less bioactive material in stem then flowers	Air dried flowers and stems	Lab
Antiviral activity	C	2016	Antiviral activity of the "Virus Blocking Factor" (VBF) derived i.a. from Pelargonium extract and Sambucus juice against different human-pathogenic cold viruses in vitro.	Fal	Minor effect	Juice	Lab
Gastrointestinal Oxidative stress	C	2016	Gastrointestinal digested Sambucus nigra L. fruit extract protects in vitro cultured human colon cells against oxidative stress.	Olejnik	Protects GI cells from Oxidative stress	Extract	Lab
Bioactive proteins	C	2016	Structural characterization of bioactive pectic polysaccharides from elderflowers (Sambuci flos).	Ho	Potential Immunomodulation region of molecule identified	Extract	Lab

APPENDIX A, CONT.

Condition	Score	Publication Year	Publication Name	Authors	Findings	Elderberry Type	Study Type
Heat Stability	C	2017	Lectin Digestibility and Stability of Elderberry Antioxidants to Heat Treatment In Vitro	Jimenez	Allergic potential reduced with brief heat treatment	Extract	Lab
Immune system stimulation	C	2016	RG-I regions from elderflower pectins substituted on GalA are strong immunomodulators.	Ho	Strong immune modulator	Extract	Lab
Differences in taste compounds	C	2016	Comparison of major taste compounds and antioxidative properties of fruits and flowers of different Sambucus species and interspecific hybrids.	Petkovsek	Significant variation between species and hybrids	Berries	Lab
Differentiation between different subspecies of Elderberry	C	2016	Variation of Select Flavonols and Chlorogenic Acid Content of Elderberry Collected Throughout the Eastern United States.	Mudge	Differentiate American elderberries from wild type	Berries	Lab
Comparison of antioxidant properties of various juices	C	2016	Multidimensional comparative analysis of phenolic compounds in organic juices with high antioxidant capacity.	Nowak	Chokeberry juices had highest concentration of antioxidant properties	Juiced fruit	Lab
Influence of ripening	C	2016	Metabolomic-Based Strategy for Fingerprinting of Sambucus nigra L. Berry Volatile Terpenoids and Norisoprenoids: Influence of Ripening and Cultivar.	Salvador	Higher metabolites in unripe fruit. Decreased as fruit ripens	Pre-harvest fruit	Lab
Bioactive content	C	2016	Effect of variety on content of bioactive phenolic compounds in common elder (Sambucus nigra L.).	Vrchotova	Less bioactive material in stem then flowers	Air dried flowers and stems	Lab
Antiviral activity	C	2016	Antiviral activity of the "Virus Blocking Factor" (VBF) derived i.a. from Pelargonium extract and Sambucus juice against different human-pathogenic cold viruses in vitro.	Fal	Minor effect	Juice	Lab
Gastrointestinal Oxidative stress	C	2016	Gastrointestinal digested Sambucus nigra L. fruit extract protects in vitro cultured human colon cells against oxidative stress.	Olejnik	Protects GI cells from Oxidative stress	Extract	Lab
Bioactive proteins	C	2016	Structural characterization of bioactive pectic polysaccharides from elderflowers (Sambuci flos).	Ho	Potential Immunomodulation region of molecule identified	Extract	Lab

APPENDIX A, CONT.

Condition	Score	Publication Year	Publication Name	Authors	Findings	Elderberry Type	Study Type
Comparison of antioxidant activities	C	2015	In vitro study of biological activities of anthocyanin-rich berry extracts on porcine intestinal epithelial cells.	Ksonzekova	Chokeberry and elderberry > blueberry or bilberry	Extract	Lab
Antioxidant activity	C	2015	Edible Flowers: A Rich Source of Phytochemicals with Antioxidant and Hypoglycemic Properties.	Loizzo	Antioxidant properties	Edible flowers	Lab
Anti-Cancer	C	2015	Aberrant glycosylation of αvβ3 integrin is associated with melanoma progression.	Pochec	Metastatic melanoma migration reduced	Sambucus nigra agglutinin	Lab
Estrogenic activity	C	2015	A molecular docking study of phytochemical estrogen mimics from dietary herbal supplements.	Powers	Docks with estrogen receptors	Dietary Supplements	Lab
Toxicity	C	2015	The Cytotoxicity of Elderberry Ribosome-Inactivating Proteins Is Not Solely Determined by Their Protein Translation Inhibition Activity.	Shang	Various mechanisms of toxicity	Protein isolates	Lab
Antioxidant/ anti-inflammatory effects	C	2015	Inhibition of microglial activation by elderberry extracts and its phenolic components.	Simonyi	Antioxidant/ anti-inflammatory effects vary with solvent used.	Extract	Lab
Comparison of Fruit Juices	C	2015	Characterization and comparison of phenolic composition, antioxidant capacity and instrumental taste profile of juices from different botanical origins.	Granato	Tests used were a suitable comparison tool	Juice	Lab
Structure and properties of proteins	C	2015	Structure-activity relationship of immunomodulating pectins from elderberries.	Ho	Structures and properties of elderberry proteins observed	Isolates	Lab
Frozen Storage	C	2015	Impact of Frozen Storage on the Anthocyanin and Polyphenol Contents of American Elderberry Fruit Juice.	Johnson	Length of frozen storage time affect properties of the Juice	Juice	Lab
Bioactivity of dried vs fresh berries	C	2015	Effect-directed analysis of fresh and dried elderberry (Sambucus nigra L.) via hyphenated planar chromatography.	Kruger	Levels of various compounds differ in dried an fresh fruits	Dried and fresh berries	Lab
Looking for Chemical Markers	C	2016	Inferring the origin of rare fruit distillates from compositional data using multivariate statistical analyses and the identification of new flavour constituents.	Mihajilov	Chemical markers found	Distillates	Lab

APPENDIX A, CONT.

Condition	Score	Publication Year	Publication Name	Authors	Findings	Elderberry Type	Study Type
Antioxidant activity	C	2015	Traditional elderflower beverages: a rich source of phenolic compounds with high antioxidant activity.	Mikulic-Petkovsek	High antioxidant activity	Extracts	Lab
Assessment of bioactive properties	C	2015	Fruit Phenolic Composition of Different Elderberry Species and Hybrids.	Mikulic-Petkovsek	Bioactive properties have practical	Extracts	Lab
Determination of adequate dosing	C	2015	Quantification of anthocyanins in elderberry and chokeberry dietary supplements.	Vlachojannis	3.5g/day influenza	Elderberry concentrate	Lab
Methods of Separating compounds	C	2015	Peptidomics study of anthocyanin-rich juice of elderberry.	Wu	1000 peptides successfully identified	Juice	Lab
Gene sequencing	C	2015	Next-generation sequencing of elite berry germplasm and data analysis using a bioinformatics pipeline for virus detection and discovery.	Ho	Gene sequencing	Berry germplasm	Lab
Toxicity	C	2015	Elderberries: a source of ribosome-inactivating proteins with lectin activity.	Tejero	Potential toxic effects	Extract	Lab
Antiviral activity	C	2014	Sambucus nigra extracts inhibit infectious bronchitis virus at an early point during replication.	Chen	Has antiviral activity	Extract	Lab
Antioxidant properties	C	2014	In vitro antioxidant properties and anthocyanin compositions of elderberry extracts.	Duymus	Infusion as potent as 70% extract	Infusion	Lab
Identification of bioactive compounds	C	2014	Screening for bioactive metabolites in plant extracts modulating glucose uptake and fat accumulation.	El-Houri	Compounds that stimulate glucose uptake identified	Extract	Lab
Anti-inflammatory activity	C	2014	Anti-inflammatory activity of fruit fractions in vitro, mediated through toll-like receptor 4 and 2 in the context of inflammatory bowel disease.	Nasef	Reduction in inflammation	Extract	Lab
Concentration of Bioactive Compounds	C	2014	Determination of ursolic and oleanolic acid in Sambuci fructus.	Glensk	High levels of bioactive compounds	Extract	Lab
Determination of elements	C	2014	Determination of elements by atomic absorption spectrometry in medicinal plants employed to alleviate common cold symptoms.	Kucukbay	Various elements identified	Elderberry Solution	Lab

APPENDIX A, CONT.

Condition	Score	Publication Year	Publication Name	Authors	Findings	Elderberry Type	Study Type
Antioxidant activities	C	2014	Polyphenols pattern and correlation with antioxidant activities of berries extracts from four different populations of Sicilian Sambucus nigra L.	Mandrone	High antioxidant potential	Berries	Lab
Bioactive Compounds	C	2014	Investigation of anthocyanin profile of four elderberry species and interspecific hybrids.	Mikulic-Petkovsek	19 different compounds detected in tested species	Extract	Lab
Heat and Digestive Stability	C	2014	Lectin Digestibility and Stability of Elderberry Antioxidants to Heat Treatment In Vitro.	De Ferrars	Heat reduces the allergic potential of Elderberry	Fruit and Bark	Lab
Analysis of carbohydrate binding properties	C	2014	Comparative analysis of carbohydrate binding properties of Sambucus nigra lectins and ribosome-inactivating proteins.	Shang	Broad range of binding capabilities	-	Lab
Glucose uptake	C	2013	Regulation of glucose transporter expression in human intestinal Caco-2 cells following exposure to an anthocyanin-rich berry extract.	Alzaid	Glucose uptake reduced	Extract	Lab
Sunscreen formulations	C	2013	Assessment of extracts of Helichrysum arenarium, Crataegus monogyna, Sambucus nigra in photoprotective UVA and UVB; photostability in cosmetic emulsions.	Jarzycka	Provide adequate levels of UV protection	Extract	Lab
Antiviral activity	C	2013	Binding of a natural anthocyanin inhibitor to influenza neuraminidase by mass spectrometry.	Swaminathan	Potential new class of antivirals	Extract	Lab
Comparison of anti-inflammatory effects of various plants	C	2013	Identification of Magnolia officinalis L. bark extract as the most potent anti-inflammatory of four plant extracts.	Walker	Magnolia officinalis L Most potent	Extract	Lab
Fatty acid content	C	2013	Lipid classes and fatty acid regiodistribution in triacylglycerols of seed oils of two Sambucus species (S. nigra L. and S. ebulus L.).	Duff	Elderberry ~44% Fatty acids in seeds	Seeds	Lab
Seed Comparison	C	2013	Comparative analyses of seeds of wild fruits of Rubus and Sambucus species from Southern Italy: fatty acid composition of the oil, total phenolic content, antioxidant and anti-inflammatory properties of the methanolic extracts.	Fazio	Seed extract have strong antioxidant properties	Extract	Lab

APPENDIX A, CONT.

Condition	Score	Publication Year	Publication Name	Authors	Findings	Elderberry Type	Study Type
Pollen proteins in Elderberry extracts	C	2013	Acyl spermidines in inflorescence extracts of elder (Sambucus nigra L., Adoxaceae) and elderflower drinks.	Kite	Elderflower drinks contain pollen proteins	Extract	Lab

*Access to abstract only

- Inaccessible or non-applicable information.

¦ Difference from Ulbricht paper.

Databases used: Pubmed.org, herbmed.org, herbalgram.org,

Can Elderberries Cure the Flu?

By Sarah Crawford, PhD

The yearly flu outbreak is going around the office. You wake up one morning with coughing, fever, muscle aches, ... oh, no! You have a big presentation due tomorrow. Is there anything you can do? You pick up the newspaper, and there it is!

Promising research suggests that extracts from the elderberry plant may help to lessen the severity and shorten the duration of the flu. Is this true, or is it just hype? Let's take a look at the evidence. The elderberry fruit has been the object of several clinical trials worldwide designed to determine whether it can alleviate the symptoms of the flu or even shorten the illness. While these studies all boast positive benefits, it is important to analyze the data with a critical eye before drawing any conclusions.

The research on the anti-viral effects of the elderberry is in two platforms: therapeutic benefit in patients sick with the flu and research on the mechanism responsible for its anti-viral effects. Both platforms need to be addressed to make a solid case for therapeutic benefit. This article focuses specifically on the perceived patient benefit; if positive results are obtained, this warrants a physiological assessment of the anti-virus components of the berries.

The Data Files

First, the good news: the consensus of clinical patient research suggests that consumption of some form of elderberry at early onset of the flu can, in some people, shorten its duration and alleviate its symptoms, including coughing, muscle aches, and fatigue. Moreover, major clinical trials evaluating the effects of elderberry in patients with the flu have generally employed a study design in which one group of patients receives the elderberry treatment and the other only a placebo.

This can be an effective way of determining whether there is any positive benefit derived from the treatment. The studies were also double-blinded, which means that neither the patients nor the researchers know which patient group receives the experimental treatment. So far, so good; however, the data are far from conclusive and fall short of suggesting that consuming elderberries should be advocated currently to treat the flu. There are several major issues that need to be addressed before any definitive conclusions can be made.

First of all, more human clinical trials are needed. So far, the anti-viral effects of this berry have been tested only in small, sporadic trials. Secondly, the size of the trial-the number of enrolled individuals- has been small in most of the reported studies. The larger the sample size, the more quantifiable the results and the greater the likelihood that the results will apply to the population at large.

For example, a clinical trial by Zakay-Rones et al. in Norway during the flu outbreak of 1999-2000 enrolled a cohort of only 60 patients (1) . Another trial by Kong in 2009 in China enrolled only 64 patients (2). A more recent clinical trial (2013-14) based in Australia involved a randomized trial of 312 study subjects (3). It is very difficult to extrapolate the responses obtained in small patient groups to wider applications.

A further issue of concern is that the elderberry supplements used in the trials were of several different types, different preparations and were combined in some cases with other supplements. For example, a clinical trial by Zakay-Rones et al. (1), in Norway during the flu outbreak of 1999-2000 used a preparation called Sambucol, an elderberry syrup. Another clinical research study conducted in China in 2009 used a proprietary (patented, secret) slow-dissolve lozenge formulation of elderberry of unknown composition (Kong) (2).

A more recent clinical trial (2013-14) based in Australia treated patients with a membrane-filtered elderberry extract (Sambucus nigra L.) that contained 300 mg of elderberry extract 15% anthocyanins and 150 mg of rice flour (3). These differences in elderberry species, concentration and the presence of additional components in some formulations make it virtually impossible to assess directly the effectiveness of the elderberry fruit in these clinical trials.

The methods used to evaluate the effectiveness of the elderberry supplements in treating the flu were also problematic. In several clinical trials, therapeutic efficacy was determined using a Visual Analog Scale (VAS) involving self-reporting by patients of the severity of their flu symptoms. Patients were asked to rate their symptoms post-treatment on a scale of 1-10, with 10 representing the highest level of discomfort.

This type of data collection is highly subjective and reported values might be susceptible to varying interpretations of the scale by individual patients. A much better approach would involve the use of objective, measurable symptom assessment by trained professionals. These evaluations might include body temperature, respiratory indices and chest X-rays where applicable. Quantitative assessments comprise a far more rigorous database that can be assessed statistically to determine the overall relevance of the findings.

Despite these study limitations, the consensus of human clinical trial data suggests that there is "good" scientific evidence that elderberry extracts lessen the severity and the duration of the flu, according to a thorough review of the anti-viral effects of the elderberry (Sambucus spp.), by Ulbricht et al. (4). Tests to measure the effects of the elderberry formulation included immune antibody production against the flu virus, clinical flu symptoms, and respiratory distress levels.

Overall, these studies indicated moderate positive effects on symptom intensity and duration as a result of elderberry treatment in patients with the flu. For example, Kong reported that 28% of the patients who took the elderberry lozenges showed complete recovery, whereas in the placebo-treated control group, no patient reported complete recovery during this time-frame(2). No adverse side effects were reported in any of the clinical trials.

Photo Credit: Shutterstock

Conclusion: Where do We Go from Here?

You may be wondering at this point when/if we will be able to say definitively whether elderberry is an effective treatment for the flu. While the clinical trial data look promising, there is still a lot that needs to be accomplished before we can answer this question with a high degree of certainty. Additional clinical trials are needed that enroll a large number of patients (as large as possible) as well as demographically diverse patient populations.

It is essential that elderberry extracts of uniform concentration, method of preparation and solvent composition be used in clinical studies designed to determine the efficacy of this therapeutic protocol. Moreover, it will be important to define dosing levels and schedules that provide the greatest therapeutic impact.

Patient responses must be evaluated by clinicians using objective response tools rather than relying on patient self-assessment to define more accurately the clinical benefit of elderberry supplements. Once these experimental criteria are met, it will be possible to determine conclusively whether elderberry extract formulations provide a beneficial effect in the treatment of flu.

REFERENCES

1. Zakay-Rones Z, Thom E, Wollan T, Wadstein J. Randomized study of the efficacy and safety of oral elderberry extract in the treatment of influenza A and B virus infections. J Int Med Res. 2004;32(2):132–140.

2. Kong F. Pilot clinical study on a proprietary elderberry extract: efficacy in addressing influenza symptoms. Online J Pharmacol Pharmacokinet. 2009;5:32–43.

3. Tiralongo, Evelin, Shirley S. Wee, and Rodney A. Lea. Elderberry supplementation reduces cold duration and symptoms in air-travellers: A randomized, double-blind placebo-controlled clinical trial. Nutrients 8.4 (2016): 182.

4. Ulbricht, Catherine, et al. An evidence-based systematic review of elderberry and elderflower (Sambucus nigra) by the Natural Standard Research Collaboration. Journal of dietary supplements 11.1 (2014): 80-120.

Which Elderberry Products are Best for the Flu?

By Sarah Crawford, PhD

> Editor's Note: The author chose to include supplements where scientific clinical data was available. We realise that a review of popular elderberry supplements might have been more helpful to the reader, but many or most of these supplements have no published clinical support.

Winter is coming soon, and, with it, another flu season. You want to start taking elderberry, but which product should you choose? I have outlined seven things to consider when purchasing a product containing elderberry. You want a product with a high concentration of the fruit, prepared and processed in a way that does not destroy its flu-fighting properties, and a cost-effective choice. Let's take a closer look to see what's out there...

Table 11 : Selected Commercially Available Elderberry Products

PRODUCT NAME/VOLUME OR WEIGHT	FORM OF ELDERBERRY	KEY ELDERBERRY FLU-FIGHTING INGREDIENTS/DOSE	RETAIL PRICE OF KEY INGREDIENTS/MG	QUALITY RANKING*
Sambucol (original)/(120ml)	Elderberry extract (38%) syrup (120ml)	3.8g of elderberry juice in each 10ml dose. 45.6g total elderberry 3800 mg=(38%)	$1.08/10ml dose; $0.28/g key elderberry	1
Mrs Miller's elderberry jelly/9 oz.	Liquid berry concentrate 25%	20g/serving (about ½ elderberry)	$2.10/ serving key elderberry; approx. $0.21 /g	2
Rubini BioFlavonoides Elderberry Extract	Elderberry concentrate (18:1) syrup	25% elderberry	N/A	3
Sambucol Extra Defense/(120ml)	Elderberry extract (38%) plus additional supplements syrup	3.8g=38%Same amount of elderberry as 'original' formula	$1.12/10ml dose; $0.30/g key elderberry/ dose	4
OptiBerry IH141/10g	Proprietary combined 6 mixed berry + vitamin C 10% capsule	Approx 7mg/ capsule = 15%	$3.18/capsule $0.45 /mg key elderberry	5
Frontier Dried Whole Elderberries/16 oz	Dried fruit, 16 ounces (453 grams)	100% berry	$1.25/ounce; $0.044/g key elderberry	6
Sinupret (Quanterra Sinus Defense)/3.9g.	Combined phytochemical tablet	36 mg =23% per tablet; 23% of the active ingredients in the tablet are elderberry (36mg per tablet)	$0.60 per tablet; $0.20/g key elderberry	7

Source: Sambucol (original) 120ml – Amazon, $12.96; Mrs. Miller's elderberry jelly, 9 Oz., $7.99 Amazon; Rubini pricing not available; Sambucol Extra Defense 120ml, $13.45 Amazon; Optiberry, 10g $27.99 Optimum Health Vitamins; Frontier Whole Elder European Berries, 16 ounces, $19.96 Amazon; Sinupret 3.9g, $14.95 Amazon.

Quality rankings- the best 1 and six lowest based on the method of preparation effects on bioactive components and elderberry concentration in the final product. These products were chosen based on several criteria: (1) Representative types of products containing elderberry; (2) Popular consumer choices for anti-flu products; (3) Proportion of the product that contains elderberry. The products are listed based on the order of "quality rankings." The data were obtained from product supplement /nutrition labels.

NA=data not available

Specific Elderberry Components with Flu-Fighting Properties

Elderberries are rich in flu-fighting bioactive compounds. Among these are flavonoids that include anthocyanins (plant pigments) and quercetin, both of which destroy free radicals. They also contain Vitamin C in large quantities (1). These ingredients have been studied individually, but may also have a combined effect when extracts prepared from whole berries are consumed.

Effect of Processing/Preparation Method on Flu-Fighting Properties of the Elderberry

Research studies have shown that products prepared from berries typically show varying degrees of loss of free radical fighting phenolics, which vary depending upon the extraction method used (2).

Typical berry processing methods involve heating the berries with sugar to make jams and jellies. This method of preparation may cause a minimal reduction in the activity of key flu-fighting molecules (3). Steam extraction to produce juices can destroy much of the anti-oxidant quercetin (4). Cold-press preparation methods appear to be less destructive to these important compounds (5).

Photo Credit: Shutterstock

What Forms of Elderberry are Commercially Available and What are the Properties of Each? (Please See Table for a Detailed List.)

Optiberry (capsules)

Optiberry contains a vegetable encapsulated mixture of six berries: blueberry, strawberry, cranberry, bilberry, raspberry, and elderberry (7). The formulation is proprietary; however, the formulation is prepared in a way that preserves the anthocyanin activity of the components. The formulation also contains Vitamin C at

10%. Assuming that elderberry comprises ⅔ of the berry component, one can estimate that 15% of the Optiberry capsule contains anthocyanins from the elderberry.

Sinupret (tablets)

Sinupret is a tablet formulation that contains a variety of plant compounds including the elderberry (36 mg) or about 23% per dose (8). Elderberry components comprise approximately 23% of each tablet.

Sambucol (original extract formulation)

Sambucol is a liquid elderberry extract (38%) containing three flavonoids at a high concentration of 3800 mg/dose (9). The children's formulation contains 19% elderberry extract (9). Sambucol, in its original syrup formulation, contains the fewest additional ingredients of the more popular elderberry products used for medicinal purposes.

Sambucol Extra Defense (liquid)

Sambucol Extra Defense is a newer, more complex formulation that contains the same concentration of elderberry extract (38%) as Sambucol original, but is supplemented with additional vitamins and minerals at fairly high concentrations. Unlike the original product, it is a multivitamin supplemented with elderberry extract.

Rubini (concentrated liquid)

Rubini is a highly concentrated elderberry extract prepared by evaporation of the water component of the fruit using a proprietary method designed to concentrate the active components of the berry (10). The elderberry-to-extract ratio of the product is 18:1, with a minimum anthocyanin concentration of 3.2%. The product, developed in 2011 by researchers in Germany, is not currently available in the U.S.

Jellies, juice

Jelly and juice are prepared by boiling extraction most commonly. Three pounds of fresh elderberries maximally yields up to three pints of juice, which is about 25% more concentrated than fresh berries. However, each serving of jelly/juice is about 50% sugar, reducing the elderberry concentration to about 62%.

Additives, Preservatives & Fillers' Side Effects

Encapsulations Optiberry contains no additives, fillers, preservatives, or artificial colors (11). At the other extreme, Sinuset contains glucose syrup, indigo carmine, aluminum hydroxide; lactose monohydrate; light magnesium oxide, shellac, silicon dioxide, titanium dioxide, and more (12). The side effects of some of these ingredients are questionable at best. Sambucol contains: glucose syrup, purified water, citric acid, and potassium sorbate (13).

Should Elderberry be a Sole Ingredient or Combined with Other Flu-Fighting Agents?

Unfortunately, not enough information is currently available to answer this question properly. The concept that "more is better" does not necessarily apply here. What is needed are more clinical trials that examine the effects of natural phytochemical extracts like the elderberry and other flu-fighting drugs and vitamins. There is some evidence, however, that the elderberry extract by itself may be as effective as other commercially available products, such as Tama-flu (14).

Cost-Effectiveness of Combined Formulations/Pure Elderberry (Vis-a-Vis Concentration of Flu-Fighting Ingredients)

The Table shows the comparative costs of some of the popular elderberry products. Sambucol products and Mrs. Miller's jelly were among the most cost-effective ($0.20-$0.30 per gram serving). The most expensive was Optiberry which cost $0.47 for the elderberry per dose, which comprised only 15% of the product. The dried elderberry tea was the most economical, priced at only $0.05 per gram.

Importance of Dose Level Standardization

Currently, there are no standard dosing guidelines established for elderberry products used in medicinal formulations. The concentrations of elderberry in the tablet, capsule and syrup formulations of the popular products ranged from 15-40%. More research is needed to determine the optimal concentrations of elderberry supplementation.

Conclusion

The Table shows "Quality Rankings" assigned based on the effects of processing on the active elderberry components. As tablets and capsules require a powdered dehydrated form of the material, these are ranked lower because this process has the greatest negative impact on the activity of the bioactive phenolics. Sambucol ranks #1 because the product is a liquid formulation with a significantly high elderberry content. However, it consists of just three flavonoids rather than the whole juice (as in jams, for example).

We still don't know how each of the ingredients of the elderberry contributes to its overall effectiveness; the entire berry may very well have greater flu-fighting capacity than selected components. The method of processing and concentrating the product Rubini is promising, but more information is needed. The jelly/juice product has the highest content of elderberry, but also contains about 50% sugar. Despite the higher sugar content, jams and jellies are ranked high (#2) because they consist of the whole berry components; the temperature used in preparation, however, may affect product contents.

The teas (using dried whole elderberries) rank low (#6) due to the inactivating effects of the drying process. There are many ways to consume your elderberries- I hope this review helps you to pick the product that's best for you!

REFERENCES

1. Porter, Randall S., and Robert F. Bode. "A Review of the Antiviral Properties of Black Elder (Sambucus nigra L.) Products." Phytotherapy Research (2017).

2. Lohachoompol, Virachnee, George Srzednicki, and John Craske. "The change of total anthocyanins in blueberries and their antioxidant effect after drying and freezing." BioMed Research International 2004.5 (2004): 248-252.

3. Amakura, Yoshiaki, et al. "Influence of jam processing on the radical scavenging activity and phenolic content in berries." Journal of Agricultural and Food Chemistry 48.12 (2000): 6292-6297.

4. Sablani, Shyam S., et al. "Effect of thermal treatments on phytochemicals in conventionally and organically grown berries." Journal of the Science of Food and Agriculture 90.5 (2010): 769-778.

5. Häkkinen, Sari H., et al. "Influence of domestic processing and storage on flavonol contents in berries." Journal of agricultural and food chemistry 48.7 (2000): 2960-2965.

6. Kwok, B. H. L., et al. "Dehydration techniques affect phytochemical contents and free radical scavenging activities of Saskatoon berries (Amelanchier alnifolia Nutt.)." Journal of food science 69.3 (2004).

7. Zafra-Stone, Shirley, et al. "Berry anthocyanins as novel antioxidants in human health and disease prevention." Molecular nutrition & food research 51.6 (2007): 675-683.

8. Ismail, C. "Pharmacology of Sinupret. Recent results on the rationale for the Sinupret compound." Hno 53 (2005): S38-42.

9. Barak, Vivian, Tal Halperin, and Inna Kalickman. "The effect of Sambucol, a black elderberry-based, natural product, on the production of human cytokines: I. Inflammatory cytokines." Eur Cytokine Netw 12.2 (2001): 290-296.

10. Veberic, R., J. Jakopic, and F. Stampar. "Flavonols and anthocyanins of elderberry fruits (Sambucus nigra L.)." II International Symposium on Human Health Effects of Fruits and Vegetables: FAVHEALTH 2007 841. 2007.

11. Ulbricht, Catherine, et al. "An evidence-based systematic review of elderberry and elderflower (Sambucus nigra) by the Natural Standard Research Collaboration." Journal of dietary supplements 11.1 (2014): 80-120.

12. Stoltenberg, Ines, et al. "PHARMACY." (2011).

13. Mumcuoglu, Madeleine. "Use of Elderberry Extract." U.S. Patent Application No. 12/159,691.

14. Elrod, Susan Marie, Phillip Greenspan, and S. Mark Tompkins. "EFFECT OF MUSCADINE CONCENTRATE AND COMMERCIAL ELDERBERRY PREPARATION ON WILD-TYPE H1N1 AND H3N2 INFLUENZA VIRUSES IN VITRO." DEVELOPMENT OF NUTRACEUTICALS AND FUNCTIONAL FOODS: ELDERBERRY (SAMBUCUS NIGRA), MUSCADINE (VITUS ROTUNDIFOLIA), AND CRAFT BEER: 100.

Elderberry v. Tamiflu: Which is Right for You?

By Sarah Crawford, PhD

Anti-viral medications such as Tamiflu (oseltamivir) are popular remedies that have been shown to decrease the duration of the illness as well as the severity of the symptoms. On the other hand, natural products from berries also possess important flu-fighting capabilities. So- what's the best berry? What's the best product to fight the flu? How do we know? Read on...

Berry v. Berry: How Do They Stack Up When it Comes to the Flu

Elderberries have the highest levels of anti-oxidants of almost all other berries (1). Among berries we commonly consume, the elderberry contains almost five times as much anti-oxidant power as strawberries; more than three times as much as raspberries, almost three times as much as blackberries and more than twice as much as blueberries (1).Researchers think that these flavonoids help to stimulate the immune system to fight infections caused by flu and other viruses (2). Moreover, the anti-inflammatory effects of the anthocyanins in elderberry provide an excellent antidote to the fever, aches, and pains that usually accompany the flu (2). Because the levels of these phytochemicals are significantly higher in elderberry than in almost all other edible berry species, this makes the elderberry the head of its class in flu-fighting ingredients.

Elderberry v. Flu Medications: How do They Work?

Tamiflu (Oseltamivir)

Oseltamivir (Tamiflu) is a very popular, although somewhat controversial anti-flu medication. It was the first oral medication approved for the treatment of the flu. The FDA approved oseltamivir in 1999 for the treatment of the flu based on positive results in ten patient clinical trials (3).

Medically, it is classified as a neuraminidase inhibitor. It is a chemical synthesized from shikimic acid derived from Chinese star anise (3). Despite the popularity of Tamiflu, there have been some problems associated with its use. For example, certain common strains of flu virus (H1N1, for example) have been shown to become resistant to its anti-viral effects (4). In addition, its use has been linked to some unpleasant side effects such as nausea, vomiting, and effects on kidney function in some individuals (5).

Elderberry

A compound called 'antivirin' identified in the black elderberry binds to flu virus particles and prevents them from entering epithelial cells of the respiratory tract to cause infection (6). In vitro studies have shown elderberry extracts to be directly effective against ten different strains of flu virus; moreover, unlike Tamiflu, viral resistance does not appear to be a problem with elderberry's use (7).

Elderberry extracts have also been shown to increase immune system responses to infections such as flu; this dual mechanism of action is a very important attribute of the elderberry (2). By both blocking the virus AND stimulating the immune system to fight the flu, there is less chance that flu virus resistance will develop as compared to drugs such as Tamiflu that attack the virus-infected cells only.

Patient clinical trials have shown that elderberry's regular use commencing at early stages of the flu shortens its duration by up to 4 days and decreases the severity of its symptoms. Antibody levels were higher in patients taking this elderberry extract demonstrating its effects on immune system activity (8). The documented effects of elderberry on immune system function provide evidence that its long-term use may have a preventive effect against infectious disease.

Elderberry v. Tamiflu: Which is Better

Unfortunately, there have not been nearly enough clinical research studies comparing the effects of elderberry extracts and Tamiflu.

One notable exception is a clinical research trial in the Czech Republic that compared the effects of Tamiflu, with an elderberry/echinacea (9) prepared as a concentrated extract of approximately 50% of each of these plant components to maximize the effects of two plant species with documented anti-flu properties.

Patients who received this preparation recovered faster with fewer side effects than those who received comparable treatment with Tamiflu.

Photo Credit: Shutterstock

Conclusion

So-which is better-Tamiflu or elderberry extract? This is how the two therapies compare in some important areas:

Mode of action

Anti-flu treatments must be initiated at the early stages of the flu; this is directly related to their modes of action. Tamiflu works by binding to virally infected cells and preventing the virus from escaping and spreading infection. Elderberry also blocks flu virus infection, but it does so by preventing the virus from entering the cell to cause infection (6,10). Both treatment modalities have shown broad-spectrum flu strain inhibition.

Side effects

Whereas no significant side effects have been associated with the use of elderberry preparations, reported side effects of Tamiflu include vomiting, diarrhea, headache and sleep issues. Rare accounts of kidney problems, seizures, and psychiatric episodes have also been reported.

Drug resistance

Resistance to flu virus strain H1N1 has been documented in Tamiflu and is a current cause of concern as this trend may spread to other flu strains treated with this product. In contrast, flu virus resistance has not been associated with the use of elderberry preparations.

Prevention

An oft-repeated proverb says "an ounce of prevention is worth a pound of cure". So true! When it comes to the elderberry, its beneficial effects on the immune system may help to prevent colds and flu.

On the other hand, there are no documented effects of Tamiflu on enhancing immune system function. Moreover, in 2017 the Centers for Disease Control stated that it does not recommend the use of Tamiflu to prevent the flu since its long-term use may cause flu virus to become resistant to its anti-flu effects.

Cost

At recommended dosages (see Table), elderberry extract use for approximately ten days costs about $15. Tamiflu is available in a generic version; in the U.S. the course of treatment costs approximately $140 in 2016 (11).

Photo Credit: Shutterstock

Table 12: Comparison of Anti-Flu Virus Remedies: Tamiflu v. Elderberry Juice

ASSESSMENT CATEGORY	TAMIFLU (OSELTAMIVIR)	ELDERBERRY JUICE*
Active ingredient	Chemical derivative of shikimic acid	Flavonoids, antivirin
Mode of action	Neuraminidase inhibitor blocks flu infection	Blocks flu infection, stimulates the immune system, attacks free radicals
Therapeutic effect	Decrease severity/ duration illness 3-4d	Decrease severity/ duration illness 3-4d
Common Side effects	Nausea, vomiting, headache	None reported
Viral resistance	Flu strain H1N1	None reported
Administration	Oral	Oral
mg/dose	150mg/day	2tsp/0.34oz/day*
Spectrum of virus strain activity	Broad spectrum	Broad spectrum
Prevention use	Not recommended	Yes, immune system activation
Approx. cost per 10 day treatment	$138*	$60***

*100% juice preparation; approx. 20g berry/oz. (8, 12)
**U.S. cost, NADAC 2016
*** River Hills Harvest elderberry juice

REFERENCES

1. United States Department of Agriculture Oxygen Radical Absorbance Capacity (ORAC) of Selected Foods – 2007

2. Barak, Vivian, Tal Halperin, and Inna Kalickman. "The effect of Sambucol, a black elderberry-based, natural product, on the production of human cytokines: I. Inflammatory cytokines." Eur Cytokine Netw 12.2 (2001): 290-296.

3. Kim, C. U. "Discovery and development of (oseltamivir, tamiflu): Rationally designed carbocyclic influenza neuraminidase inhibitor." ABSTRACTS OF PAPERS OF THE AMERICAN CHEMICAL SOCIETY. Vol. 219. 1155 16TH ST, NW, WASHINGTON, DC 20036 USA: AMER CHEMICAL SOC, 2000.

4. Lackenby, A., et al. "Emergence of resistance to oseltamivir among influenza A (H1N1) viruses in Europe." Eurosurveillance13.5 (2008): 3-4.

5. Fuyuno, Ichiko. "Tamiflu side effects come under scrutiny." (2007): 358-359.

6. Varshney, Jonish, et al. "VARIOUS REMEDIES FOR SWINE FLU."Pharmacologyonline 2: 36-47 (2010).

7. Krawitz, Christian, et al. "Inhibitory activity of a standardized elderberry liquid extract against clinically-relevant human respiratory bacterial pathogens and influenza A and B viruses." BMC complementary and alternative medicine 11.1 (2011): 16.

8. Zakay-Rones, Z., et al. "Randomized study of the efficacy and safety of oral elderberry extract in the treatment of influenza A and B virus infections." Journal of International Medical Research32.2 (2004): 132-140.

9. Rauš, Karel, et al. "Effect of an Echinacea-based hot drink versus oseltamivir in influenza treatment: a randomized, double-blind, double-dummy, multicenter, noninferiority clinical trial." Current Therapeutic Research 77 (2015): 66-72.

10. Roschek, Bill, et al. "Elderberry flavonoids bind to and prevent H1N1 infection in vitro." Phytochemistry 7010 (2009): 1255-1261.

11. NADAC 12-7-2016.

12. Vlachojannis, J. E., Melainie Cameron, and Sigrun Chrubasik. "A systematic review on the sambuci fructus effect and efficacy profiles." Phytotherapy Research 24.1 (2010): 1- 8.

Overview of EU Regulation of Elderberry as Medicine

By Nnenna Udensi, MD

Considered on its medicinal merits, elderberry is currently viewed as a herbal supplement by much of the world. Different countries have varying regulations for vitamins and supplements and governments are interested in how companies market health products to the public. Most governments regulate the types of claims that companies make to the public about a supplements' effects based on the quantity and quality of the data that back those claims.

In the United States, this role is fulfilled by the Food and Drug Administration (FDA). But in this article I will focus on the European Medicines Agency (EMA) which is the European Union agency responsible for the regulation of elderberry and other medicinal products in the EU. This article will touch on the herbal medicine branch of the EMA, elderberry's classification as a medicine in the EU and supporting documents such as the assessment report (EPAR) and EU Monograph for elderberry that are produced by the agency.

Overview of Herbal Medicine Regulatory Designations in the EU

The European Medicines Agency (EMA) is the organization responsible for the evaluation of medicinal products in the European Union. When drug companies or researchers want to sell their medications in EU countries, they must submit an application and seek approval from the EMA.

There is a branch of the EMA that deals specifically with the approval and classification of herbal medicinal products, such as elderberry, called the Committee for Herbal Medicinal Products (HMPC). Herbal Medicine (medicinal preparations derived entirely from plant sources) is generally considered lower risk and is not subject to the same stringent testing and guidelines that more modern therapies are.

Accordingly, this separate EMA branch for herbal products uses a simplified registration procedure. Herbal medicines are classified under either 'traditional use' or 'well-established use' as follows:

Traditional use registration

- Intended for herbal products with a long (30+ years) history of medicinal use that falls short of requirements for "well-established use."
- Safety data is sufficient and "plausible" and efficacy is demonstrated. No clinical tests or trials are required for this classification.
- Medical products in this category are intended for use without medical supervision and are not injectable.
- Must have been in use for at least 30 years including 15 years in the EU.

Well – established use marketing authorization

- Assessment of safety and efficacy data
- At least ten years of scientific literature confirming the product's active ingredients have been in well-established medicinal use.

Stand-alone/ mixed application

- Safety and efficacy data from the company's product development.

Photo Credit: Shutterstock

Elderberry Approved as a Medicinal Product Under "Traditional Use"

Elderberry is currently approved under the 'traditional use' registration which makes sense for several reasons including 1) it has been around as a treatment in numerous cultures for centuries, 2) there are clinical trials and a wealth of data establishing its safety, and 3) the efficacy for colds and other viral ailments has been studied repeatedly.

I believe elderberry might be a good candidate for a 'well-established' use marketing authorization if it was more widely used. Unfortunately, while the data clearly establishes that elderberry has some benefit in various conditions, particularly in fighting colds, the magnitude of those benefits is still debatable(8,9). Consequently, I believe the medical community tends to proceed with products that have stronger supporting data. That said, the current classification of 'traditional use registration' acknowledges elderberry's potential benefits and allows for it to be consumed as a supplement without medical supervision and to be marketed in the EU as such.

Classification of elderberry under well-established use would bring it into the doctor's office as potential treatments for approved conditions. Companies that make elderberry products would be able to market them to doctors and patients. And similar to the way a patient with a cold may be advised to take vitamin c, or someone with a wound might be prescribed zinc supplements; elderberry products would potentially be recommended by doctors with greater frequency, increasing both awareness and data available to be studied.

Monographs & Assessment Reports are Helpful to Researchers

Apart from its role in classifying medicines, the EMA also maintains a monograph of each drug, and publishes an assessment report. A monograph is a detailed study of a specialized subject and serves as a central source of information on the topic.

The EMA Monograph offers a list of herbal substances, preparations, and combinations that can be used as reference material when applying for drug registration. The EMA writes a European Public Assessment Report (EPAR) for every medicine, including elderberry, that is granted marketing by the EU. It serves as a "public-friendly" overview on the medicine(10) and is made available to the public.

The HMPC considers the need for review and revision of herbal medications every five years from the date of latest publication of the EPAR. This ensures that all the research and data is up to date. However, even when the information is updated, I am unclear if, absent another application, the classification is automatically upgraded to 'well-established use' marketing due to new studies. The last assessment report for elderberry was published in 2013. So perhaps an update is in the works this year, although I could not find any evidence of one. The EU Monograph for elderberry was updated in March 2018. An updated assessment report (EPAR) or even just the updated European Union Monograph for elderberry is valuable to all the stakeholders who are interested in elderberry's use in medicine. Worldwide, there have been 100+ studies on elderberry in the past five years on a range of topics.

While not all of the new data is conclusive, quick and easy access to quality, timely information allows medical providers who recommend supplements to their patients, or lay people looking for useful information on medicinal products, to make more educated decisions. And for individuals or companies who may choose to develop medications further, the monograph information can facilitate their research.

Questions about quality, non-clinical, clinical, and safety concerns are addressed in the EPAR. The EPAR (assessment report) includes:

Quality guidelines
- Description of herbal substance, preparation, and characteristics.
- Quality starting materials, good agricultural and collection practices
- Quality of product combinations
- Testing procedures and acceptance criteria for medicinal products

Non-clinical
- Assessment of toxicity of preparations
- Medical applications
- Chemistry data
- Historical data

Clinical
- Safety and efficacy (does it work?)
- Clinical evaluation of fixed combinations
- Dosage

Safety
- Available information on toxicology
- Serious adverse events
- Special populations (children, pregnancy, etc.)

As part of the application process, the HMPC also issues a report that reviews statements about the product that are given by the public (public comments). Absent rigorous studies about the particular herbal medicine being evaluated, it seems that the HMPC considers public consultation a valuable resource and uses it to bring to light potential benefits, or issues not addressed in the original application documents.

Any concerns can then be further vetted. The comments during the 2013 application process for elderberry addressed dosing, preparation, the 30+ years of medicinal use, conditions, and also allowed for application authors to answer comments.

Conclusions

The EMA and FDA and similar governmental organizations around the world seek to protect consumers in many ways, including, requiring good clinical data for claims made about products, reviewing that data, making the data accessible to the medical and scientific communities, and regulating drugs and products marketed to the public.

The strictest regulations are usually applied to newly developed drugs and tend to be less stringent for herbal substances like elderberry. Elderberry is an active area of medical research in the EU and worldwide, and the Committee for Herbal Medicinal Products (HMPC) updated its Monograph earlier this year. But given that elderberry remains in limited use by medical professionals in the EU, I doubt elderberry will be upgraded to well-established use for marketing purposes in the EU anytime soon. Perhaps data from new studies or more ambitious studies concerning elderberry's efficacy will spur such a change in the future. For now, elderberry will continue to be used as an over the counter supplement to remedy various ailments.

REFERENCES

1.   Assessment Report On Sambucus Nigra L., Fructus. European Union: Committee on Herbal Medicinal Products (HMPC); 2013:1-26. http://www.ema.europa.eu/docs/en_GB/document_library/Herbal_-_HMPC_assessment_report/2013/04/WC500142245.pdf. Accessed September 24, 2018.

2.   Procedure For The Review And Revision Of European Union 4 Herbal Monographs And European Union List Entries. 2nd ed. European Union: Committee on Herbal Medicinal Products (HMPC); 2017:1-15. http://www.ema.europa.eu/docs/en_GB/document_library/Scientific_guideline/2017/10/WC500237848.pdf. Accessed September 24, 2018.

3.   Sambucus Nigra L., Fructus (EMA/HMPC/44208/2012). European Union: Committee on Herbal Medicinal Products (HMPC); 2014:1-24. http://www.ema.europa.eu/docs/en_GB/document_library/Herbal_-_Overview_of_comments_received_during_consultation/2014/04/WC500165896.pdf. Accessed September 25, 2018.

4.   Elderberry Juice - Biotta Juices. Biotta Juices. http://www.biottajuices.com/products/elderberry-juice/. Published 2018. Accessed September 25, 2018.

5.   Sambucus nigra (Elderberry). Alternative Medicine Review. 2005;10(1):1-55. http://www.altmedrev.com/archive/publications/10/1/51.pdf. Accessed September 25, 2018.

6.   ELDERBERRY Monograph: Natural Medicines Comprehensive Database. Naturaldatabase.therapeuticresearch.com. http://naturaldatabase.therapeuticresearch.com/nd/Search.aspx?cs=&s=ND&pt=100&id=434&ds=. Published 2018. Accessed September 25, 2018.

7.   American Botanical Council: Herb Med Pro. Cms.herbalgram.org. http://cms.herbalgram.org/herbmedpro/index.html?ts=1537745673&signature=e80f61dcb273ff3cc9c902e1d64528df#param.wapp? sw_page=@@viewHerb%3FherbID%3D64. Published 2018. Accessed September 25, 2018.

8.   Zakay-Rones Z, Thom E, Wollan T, Wadstein J. Randomized Study of the Efficacy and Safety of Oral Elderberry Extract in the Treatment of Influenza A and B Virus Infections. Journal of International Medical Research. 2004;32(2):132-140. doi:10.1177/147323000403200205.

9. Tiralongo E, Wee SS, Lea RA. Elderberry Supplementation Reduces Cold Duration and Symptoms in Air-Travellers: A Randomized, Double-Blind Placebo-Controlled Clinical Trial. Nutrients. 2016;8(4):182. doi:10.3390/nu8040182.

10. European public assessment reports. European Medicines Agency - Find medicine - Raxone. https://www.ema.europa.eu/about-us/how-we-work/what-we-publish/european-public- assessment-reports. Accessed September 29, 2018.

APPENDIX B

SUMMARY OF EUROPEAN MEDICINES AGENCY ASSESSMENT REPORT ON SAMBUCUS NIGRA. (ELDERBERRY)

Non-Clinical Data	
Elderberry Activities	**Findings**
Immunological Activity	• May modulate activity of immune cells. • Data is not conclusive.
Antiviral Activity	• Inhibitory effect on influenza virus strains studied. • Clear results in lab studies • Mixed results in studies on human cells*.
Antioxidative activity	• High concentrations of chemicals known for antioxidant properties. • Mixed study results on effects of stress in cells* in vitro. • Applications Unclear
Antibacterial activity	• Antibacterial activity found against H. Pylori and MRSA.
Lectin Activities	• Lectins are sugar binding proteins • Lectin activities have been established by studies • Applications for lectin activities (diabetes) has been studied. Mixed results.
Toxicology	• High-risk of toxicity with ripe fruit • No concerns with cooked or heat treated products. • Avoid bark of the plant. (toxic)

APPENDIX B, CONT.

Clinical Data	
Elderberry Activities	**Findings**
Low Bioavailability	• The proportion of active substances in elderberry that enter the circulation after being consumed is low across all preparations.
Dose response	• Little data is available for dosing, likely due to low bioavailability.

Safety Data	
Elderberry Activities	**Findings**
Adverse Events	• Adverse events in a few individuals have been reported. • Abdominal pain, Edema in Legs, Decreased INR (increased blood clotting)

For the purposes of this report, Non clinical data represents pharmacodynamics, and toxicology from laboratory (in-vitro) experiments. Clinical data represents pharmacokinetic, bioavailability, and clinical trial data. Safety data represents information on clinical exposures.

REFERENCES TO THE APPENDIX

1. Assessment Report On Sambucus Nigra L., Fructus. European Union: Committee on Herbal Medicinal Products (HMPC); 2013:1-26. http://www.ema.europa.eu/docs/en_GB/document_library/Herbal_-_HMPC_assessment_report/2013/04/WC500142245.pdf. Accessed September 24, 2018.

2. Procedure For The Review And Revision Of European Union 4 Herbal Monographs And European Union List Entries. 2nd ed. European Union: Committee on Herbal Medicinal Products (HMPC); 2017:1-15. http://www.ema.europa.eu/docs/en_GB/document_library/Scientific_guideline/2017/10/WC500237848.pdf. Accessed September 24, 2018.

3. Sambucus Nigra L., Fructus (EMA/HMPC/44208/2012). European Union: Committee on Herbal Medicinal Products (HMPC); 2014:1-24. http://www.ema.europa.eu/docs/en_GB/document_library/Herbal_-_Overview_of_comments_received_during_consultation/2014/04/WC500165896.pdf. Accessed September 25, 2018.

4. Elderberry Juice - Biotta Juices. Biotta Juices. http://www.biottajuices.com/products/elderberry-juice/. Published 2018. Accessed September 25, 2018.

5. Sambucus nigra (Elderberry). Alternative Medicine Review. 2005;10(1):1-55. http://www.altmedrev.com/archive/publications/10/1/51.pdf. Accessed September 25, 2018.

6. ELDERBERRY Monograph: Natural Medicines Comprehensive Database. Naturaldatabase.therapeuticresearch.com. http://naturaldatabase.therapeuticresearch.com/nd/Search.aspx?cs=&s=ND&pt=100&id=434&ds=. Published 2018. Accessed September 25, 2018.

7. American Botanical Council: Herb Med Pro. Cms.herbalgram.org. http://cms.herbalgram.org/herbmedpro/index.html?ts=1537745673&signature=e80f61dcb273ff3cc9c902e1d64528df#param.wapp? sw_page=@@viewHerb%3FherbID%3D64. Published 2018. Accessed September 25, 2018.

Section 4:
Special
Considerations

How Much Elderberry Should I Take? What Should the Dosage be?

By Nnenna Udensi, MD

People around the world use elderberry in various foods including wine, jams, desserts, drinks, etc., and when used for these purposes, elderberry is added for taste or nutrition. For medicinal preparations though, ideally, medical professionals need to know how much of the active ingredients should be administered to achieve a particular effect.

But current supplemental preparations containing elderberry vary in the concentration of active elderberry ingredients present because there is currently no standardization in dosing of elderberry for specific diseases. Furthermore, given that the effectiveness of elderberry for treating disease has not been proven, it is difficult to speak to the correct dose of elderberry for given conditions.

However, what I can speak to is the amounts given in clinical trials I reviewed, some of which have statistically significant results, and the amounts that can be safely consumed. Please keep in mind that these doses (below) have not been standardized and cannot be used as a guide for consumers who wish to take elderberry as a supplement.

In recent clinical trials(9,11,12,13) (2015-2018) the doses tested ranged from 300mg/day – 500mg/day, and the condition most commonly tested was influenza infection. The periods tested varied from one day(9) to five weeks(13).

An over the counter lozenge preparation that contains 175mg of extract and is taken 4/day for a total of 700mg daily for 2 days(9).

Certain over the counter syrup preparations recommend 15ml 4 times daily(8). A juice preparation that was studied was the juice of 50 g of elderberries over 5 weeks(13).

In the forms that are typically consumed as supplements: fruit syrups, powdered or liquid extracts, and juices, the range of doses varies. However, on average, the concentrations of syrups range from 30-40% elderberry, the extracts average in 500mg 2-4 times daily range, and the Juices vary based on concentration(1,4,6,9). The wide range in dosage recommendations by different companies is indicative of the lack of research in this area for elderberry.

If you're confused about how much elderberry you should take for some expected medical benefit, you should be! The medical data is just not there. As far as evidence-based dosing for over the counter preparations, I wasn't able to find any. While small doses where tested in clinical trials, I was unable to find any data on a maximum safe dose. But given the strong safety data for correct preparations of elderberry combined with the relatively low use in modern medicine, it likely has not been explicitly studied.

Overall, more studies on dosage will help to standardize the preparations in the market. Alongside testing the effectiveness of elderberry against different disease states, going forward, studies will also hopefully test specifically how much or how little elderberry it takes to achieve those effects. Until then, elderberry products will continue to reflect the doses studied or the approximations of the companies that sell them.

Photo Credit: Shutterstock

REFERENCES

1. Assessment Report On Sambucus Nigra L., Fructus. European Union: Committee on Herbal Medicinal Products (HMPC); 2013:1-26. http://www.ema.europa.eu/docs/en_GB/document_library/Herbal_-_HMPC_assessment_report/2013/04/WC500142245.pdf. Accessed September 24, 2018.

2. Procedure For The Review And Revision Of European Union 4 Herbal Monographs And European Union List Entries. 2nd ed. European Union: Committee on Herbal Medicinal Products (HMPC); 2017:1-15. http://www.ema.europa.eu/docs/en_GB/document_library/Scientific_guideline/2017/10/WC500237848.pdf. Accessed September 24, 2018.

3. Sambucus Nigra L., Fructus (EMA/HMPC/44208/2012). European Union: Committee on Herbal Medicinal Products (HMPC); 2014:1-24. http://www.ema.europa.eu/docs/en_GB/document_library/Herbal_-_Overview_of_comments_received_during_consultation/2014/04/WC500165896.pdf. Accessed September 25, 2018.

4. Elderberry Juice - Biotta Juices. Biotta Juices. http://www.biottajuices.com/products/elderberry-juice/. Published 2018. Accessed September 25, 2018.

5. Sambucus nigra (Elderberry). Alternative Medicine Review. 2005;10(1):1-55. http://www.altmedrev.com/archive/publications/10/1/51.pdf. Accessed September 25, 2018.

6. ELDERBERRY Monograph: Natural Medicines Comprehensive Database. Naturaldatabase.therapeuticresearch.com. http://naturaldatabase.therapeuticresearch.com/nd/Search.aspx?cs=&s=ND&pt=100&id=434&ds=. Published 2018. Accessed September 25, 2018.

7. American Botanical Council: Herb Med Pro. Cms.herbalgram.org. http://cms.herbalgram.org/herbmedpro/index.html?ts=1537745673&signature=e80f61dcb273ff3cc9c902e1d64528df#param.wapp? sw_page=@@viewHerb%3FherbID%3D64. Published 2018. Accessed September 25, 2018.

8. Zakay-Rones Z, Thom E, Wollan T, Wadstein J. Randomized Study of the Efficacy and Safety of Oral Elderberry Extract in the Treatment of Influenza A and B Virus Infections. Journal of International Medical Research. 2004;32(2):132-140. doi:10.1177/147323000403200205.

9.   Tiralongo E, Wee SS, Lea RA. Elderberry Supplementation Reduces Cold Duration and Symptoms in Air-Travellers: A Randomized, Double-Blind Placebo-Controlled Clinical Trial. Nutrients. 2016;8(4):182. doi:10.3390/nu8040182.

10. European public assessment reports. European Medicines Agency - Find medicine - Raxone. https://www.ema.europa.eu/about-us/how-we-work/what-we-publish/european-public- assessment-reports. Accessed September 29, 2018.

11. Levine WZ, Samuels N, Sheshet MEB, Grbic JT. A Novel Treatment of Gingival Recession using a Botanical Topical Gingival Patch and Mouthrinse. The Journal of Contemporary Dental Practice. 2013;14:948-953. doi:10.5005/jp-journals-10024-1431.

12. Vauzour D, Tejera N, Oneill C, et al. Anthocyanins do not influence long-chain n-3 fatty acid status: studies in cells, rodents and humans. The Journal of Nutritional Biochemistry. 2015;26(3):211-218. doi:10.1016/j.jnutbio.2014.09.005.

13. Nilsson A, Salo I, Plaza M, Björck I. Effects of a mixed berry beverage on cognitive functions and cardiometabolic risk markers; A randomized cross-over study in healthy older adults. PLoS ONE. 2017;12(11):e0188173. doi:10.1371/journal.pone.0188173

Current Scientific Understanding of the Side Effects and Interactions from Consuming Elderberry

By Nnenna Udensi, MD

For centuries, traditional medicine in a myriad of cultures has used both wild and cultivated elderberry to treat and prevent ailments. Given its numerous uses over such a long period of time, it's no wonder that the value of elderberry is still recognized generations later and now increasingly within the medical community.

Medicinal applications aside, elderberry is often consumed simply in its capacity as fruit and in meal preparations derived from fruit such as desert and juices. That said, as with many other plant-derived foods in our diet, there is always the possibility of unwanted reactions or interactions.

This report provides an overview of potential undesired interactions from consuming elderberry. The majority of the data for elderberry concerning precautions, adverse events, and contraindication revolves around the toxicity of raw berries, bark and other parts of the plant; the resulting gastric distress, and use in pregnant or lactating women.

Potential medication and allergic interactions are also described. Otherwise, I did not find any specific contraindications to elderberry use described in the literature. To prepare this report I reviewed scientific articles relating to this subject from Google Scholar and Pub Med covering the time period 1982 to 2018.

Photo Credit: Shutterstock

Potential for Interactions and Side Effects - Hydrogen Cyanide (HCN)

One particularly harmful aspect is that consuming raw elderberries is toxic(1). The ripe fruit with its high sugar content can also produce hydrogen cyanide while undergoing fermentation. It seems that as the fruit progresses through ripening, fermentation, and decay, the concentration of hydrogen cyanide and its substrates increase(13).

Hydrogen cyanide(11,13) is a poison that affects cell biology by preventing the cell from utilizing oxygen. It causes headaches, fast heart rate, vomiting, and diarrhea initially; and prolonged or substantial exposure can cause seizures and low blood pressure(8,17). Fortunately, heat treatment(7) of the ripe, raw fruit at 150 degrees for about 30 minutes reduces the hydrogen cyanide without destroying the compounds responsible for the potential health benefits in both ripe and fermented berries(13).

So it is recommended that elderberry fruit be "cooked" before consumption. It is important to note that this process stops well short of boiling, which would render the beneficial properties of elderberry ineffective(7).

There are also reports(21) of unripe elderberries causing gastric distress in young children, however the vast majority of the literature is devoted to the effects of ripe fruit. Elderberry is not unique when it comes to this food toxicity. Raw bitter almonds are notorious for producing the same toxin and are largely banned in the US. Toxicity levels also depend on the species of elderberry, some being far more toxic than others.

Potential for Interactions and Side Effects - Ribosome Inactivating Proteins (RIP)

The seeds, leaves, stem, bark, roots of the Elder plant are potentially harmful as well, especially if infused and consumed as an herbal tea(1). In addition to hydrogen cyanide, these parts contain chemical substances called ribosome inactivating proteins (RIP), that affect cell biology by inactivating the structures in animal cells that create proteins(1,2).

Not all RIPs are toxic to humans when consumed. They exist in plants and bacteria as a form of protection(15). Elderberry's RIPs are toxic to insects prompting research into its potential use as an insecticide. However, despite the clear demonstration of toxicity to human cells(16), the effects studied so far are limited mostly to laboratory experiments.

Potential for Interactions and Side Effects - Allergies

As with any food, allergies should be considered when consuming it for the first time. That includes elderberry, where the risk of allergy is likely to be higher in people who are also allergic to honeysuckle as the plants are closely related. It seems the RIPs present in elderberry have allergenic potential that can be reduced to some degree with heat treatment(7).

Potential for Interactions and Side Effects - When Taken with Other Medicine

Thus far I have reviewed the potential risks from consuming elderberry as food or drink. Taken medicinally, however, the potential side effects and interactions of elderberry are less clear. While there are no confirmed drug interactions with elderberry, there is a slew of theoretical and laboratory identified interactions.

The distinction between theory and actual lab results is important because although the theories have not been empirically tested, the information can still be considered when doing risk assessments for a particular disease or therapy. For instance, given the reports of elderberry's laxative and diuretic effects(17,22), someone already taking medications for constipation or diuresis may want to avoid the addition of elderberry to their regimen as excessive diarrhea and diuresis can lead to electrolyte imbalances and other undesired effects. There are other potential effects as well.

Currently research suggests caution should be exercised when consuming an elderberry supplement together with:

Antidiabetic drugs(4)
Active ingredients in elderberry have documented effects on binding of sugar in the blood. Potential for hypoglycemia if combined with drugs with similar effects.

Morphine(6)

An animal study showed decreased analgesic action after the subject also consumed elderberry. Potential for reducing the effectiveness of pain medications

Phenobarbital(6)

An animal study showed a decrease in induction time or the time it takes to have its effect, an increase in sleep time. This potential effect may be acceptable if the person taking the medications has trouble falling and staying asleep. However, it may be undesired in a person undergoing anesthesia.

Immuno-active drugs(5,10,12)

Several laboratory and animal studies have shown that chemicals present in elderberry interact with the immune system in various ways. So there is an argument to be made that other medications designed to interact with the immune system may also interact with elderberry.

Caution is also advised in certain groups due to higher vulnerability, and lack of explicit safety data in these groups. The medical field generally avoids taking unnecessary risks in pregnant and lactating women, young children, and people with compromised immune systems.

Potential for Interactions and Side Effects - Immunocompromised Individuals

The literature on side effects in this patient population is a bit more nuanced. The potential for small beneficial outcomes, often involving symptom relief, is described in the literature(12). However, focusing on short-term gains in this way is short sighted. Particularly when one takes into account elderberry's ability to alter normal functioning of the immune system in ways yet unclear to us(12).

In theory, elderberry could affect autoimmune disease precipitated inflammation, thus calming symptoms and potentially benefiting immune conditions such as arthritis, diabetes, or inflammatory bowel disease. Several studies(10) show elderberry to reduce inflammation in laboratory and animal studies. But this effect has not yet been studied as extensively in autoimmune disease or clinical trials.

Conversely, given studies describing immune modulating effects(5,12), treatment could also further modify an already malfunctioning immune system in ways we don't yet understand. In the absence of conclusive evidence reporting the benefits of elderberry treatment in these diseases, it is difficult to justify exposing affected patient populations to the potential risks. Especially when proven and cost-effective therapies are available.

The appeal is understandable though. Immunocompromised disease states are especially pertinent because they often involve lifelong treatment with potent, and potentially expensive drugs. These drugs can be difficult to take or cause side effects of their own, sending patients looking for answers and relief wherever they can find it.

This makes the lure of alternative therapies strong for some looking for natural ways to treat lifelong illnesses. And for some immune diseases, the data in support of elderberry having benefits for these illnesses shows potential.(5,12) Even still, this population of patients should exercise caution. Effects tolerated by people with healthy immune systems may vary in people with immune disorders. The only thing current research in this area tells us conclusively is that we do not know.

In Conclusion

After reviewing available scientific research on elderberry, I would describe it as a useful supplement that has an array of potentially beneficial effects. The better documented and studied benefits primarily involve the antiviral/microbial, antioxidant, and anti-inflammatory effects of elderberry.

With these effects as mechanisms of action, elderberry seems to potentially affect a wide range of conditions including influenza(27,28), hyperlipidemia(23,24) obesity(29), gingivitis(25,26), stress(23), diabetes(4), inflammatory diseases (23,30) and more. When prepared correctly, research shows that consumption of elderberry is safe in several different forms and doses, over various periods of time.

As a clinician, I would advise people considering medicinal elderberry use to ensure it is prepared correctly or otherwise stick to supplement preparations. There are several different preparations to choose from, but extracts have more robust data.

Start with a small dose to test for allergic reactions and stay within the manufacturers' recommended dosage for the product. Nevertheless, given the wide range of dosages indicated by supplement manufacturers, it's unclear how they are establishing their recommendations.

I would suggest limiting use of elderberry supplements to the duration of the symptoms for colds and upper-respiratory infections and discourage use for chronic conditions where proven medications are readily available.

For supplemental use it is safe, and perhaps a good idea in adults whose diet is lacking similar berries and fruits. There is little data on long-term use, but likely safe in moderation over extended periods of time. I would caution against pregnant/nursing women, small children, and immunocompromised people using this supplementation since safety studies have not been be done in these groups, and the benefits may not outweigh the potential risks.

A diet higher in dark berries and other similar fruits would provide a comparable level of nutrition and benefit in these groups and should be pursued first. While elderberry's efficacy continues to be studied and debated in academia, the increasing number of studies seeking to answer questions regarding its additional medical potential, are one of the highest forms of validation the medical field can offer.

It suggests that governments, scientists, academics, doctors, and pharmaceutical companies all recognize the medical potential of elderberry and think it a worthwhile pursuit. The scientific community is gradually coming to accept some of the beliefs and anecdotes about the benefits of elderberry held by many civilizations that predate ours.

Photo Credit: Shutterstock

REFERENCES

1. Battelli MG, Citores L, Buonamici L, et al. Toxicity and cytotoxicity of nigrin b, a two-chain ribosome-inactivating protein from Sambucus nigra : comparison with ricin. Archives of Toxicology. 1997;71(6):360-364. doi:10.1007/s002040050399

2. Bratu MM, Doroftei E, Negreanu-Pirjol T, Hostina C, Porta S. Determination of Antioxidant Activity and Toxicity of Sambucus nigra Fruit Extract Using Alternative Methods. April 2012:1-6.

3. Choi EJ, Park JB, Yoon KD, Bae SK. Evaluation of the in vitro/in vivo potential of five berries (bilberry, blueberry, cranberry, elderberry, and raspberry ketones) commonly used as herbal supplements to inhibit uridine diphospho-glucuronosyltransferase. Food and Chemical Toxicology. 2014;72:13-19. doi:10.1016/j.fct.2014.06.020

4. Ciocoiu M, Mirón A, Mares L, et al. The effects of Sambucus nigra polyphenols on oxidative stress and metabolic disorders in experimental diabetes mellitus. Journal of Physiology and Biochemistry. 2009;65(3):297-304. doi:10.1007/bf03180582

5. Citores L, Iglesias R, Muñoz R, Ferreras J, Jimenez P, Girbes T. Elderberry (Sambucus nigraL.) seed proteins inhibit protein synthesis and display strong immunoreactivity with rabbit polyclonal antibodies raised against the type 2 ribosome-inactivating protein nigrin b. Journal of Experimental Botany. 1994;45(4):513-516. doi:10.1093/jxb/45.4.513

6. Jakovljevi V, Popovic M, Mimica-Dukic N, Sabo J. Interaction of Sambucus nigra Flower and Berry Decoctions with the Actions of Centrally Acting Drugs in Rats. Pharmaceutical Biology. 2001;39(2):142-145. doi:10.1076/phbi.39.2.142.6260

7. Jiménez P, Cabrero P, Cordoba-Diaz D, Cordoba-Diaz M, Garrosa M, Girbés T. Lectin Digestibility and Stability of Elderberry Antioxidants to Heat Treatment In Vitro. Molecules. 2017;22(1):95. doi:10.3390/molecules22010095

8. Leads from the MMWR. Poisoning from elderberry juice. JAMA: The Journal of the American Medical Association. 1984;251(16):2075-2075. doi:10.1001/jama.251.16.2075

9. Mccune JS, Hatfield AJ, Blackburn AAR, Leith PO, Livingston RB, Ellis GK. Potential of chemotherapy?herb interactions in adult cancer patients. Supportive Care in Cancer. 2004;12(6):454-462. doi:10.1007/s00520-004-0598-1

10. Millar CL, Norris GH, Jiang C, et al. Long-Term Supplementation of Black Elderberry Promotes Hyperlipidemia, but Reduces Liver Inflammation and Improves HDL Function and Atherosclerotic Plaque Stability in Apolipoprotein E-Knockout Mice. Molecular Nutrition & Food Research. 2018:1800404. doi:10.1002/mnfr.201800404

11. National Research Council (US) Subcommittee on Acute Exposure Guideline Levels. Hydrogen Cyanide: Acute Exposure Guideline Levels. Acute Exposure Guideline Levels for Selected Airborne Chemicals: Volume 2. https://www.ncbi.nlm.nih.gov/books/NBK207601/. Published January 1, 1970. Accessed October 14, 2018.

12. Newkirk MM, Fournier M-JE, Shiroky J. Rheumatoid factor avidity in patients with rheumatoid arthritis: Identification of pathogenic RFs which correlate with disease parameters and with the gal(O) glycoform of IgG. Journal of Clinical Immunology. 1995;15(5):250-257. doi:10.1007/bf01540882

13. Pogorzelski E. Formation of cyanide as a product of decomposition of cyanogenic glucosides in the treatment of elderberry fruit (Sambucus nigra). Journal of the Science of Food and Agriculture. 1982;33(5):496-498. doi:10.1002/jsfa.2740330516

14. Senica M, Stampar F, Veberic R, Mikulic-Petkovsek M. Processed elderberry (Sambucus nigra L.) products: A beneficial or harmful food alternative? LWT - Food Science and Technology. 2016;72:182-188. doi:10.1016/j.lwt.2016.04.056

15. Shahidi-Noghabi S, Damme EJMV, Smagghe G. Expression of Sambucus nigra agglutinin (SNA-I☐) from elderberry bark in transgenic tobacco plants results in enhanced resistance to different insect species. Transgenic Research. 2008;18(2):249-259. doi:10.1007/s11248-008- 9215-2

16. Shang C, Chen Q, Dell A, Haslam SM, Vos WHD, Damme EJMV. The Cytotoxicity of Elderberry Ribosome-Inactivating Proteins Is Not Solely Determined by Their Protein Translation Inhibition Activity. Plos One. 2015;10(7). doi:10.1371/journal.pone.0132389

17. Sidor A, Gramza-Michałowska A. Advanced research on the antioxidant and health benefit of elderberry (Sambucus nigra) in food – a review. Journal of Functional Foods. 2015;18:941-958. doi:10.1016/j.jff.2014.07.012

18. Sprouse AA, Breemen RBV. Pharmacokinetic Interactions between Drugs and Botanical Dietary Supplements. Drug Metabolism and Disposition. 2015;44(2):162-171. doi:10.1124/dmd.115.066902

19. Williamson EM. Drug Interactions Between Herbal and Prescription Medicines. Drug Safety. 2003;26(15):1075-1092. doi:10.2165/00002018-200326150-00002

20. Zheng Y, Min J, Kim D, et al. In Vitro Inhibition of Human UDP-Glucuronosyl-Transferase (UGT) Isoforms by Astaxanthin, ?-Cryptoxanthin, Canthaxanthin, Lutein, and Zeaxanthin: Prediction of in Vivo Dietary Supplement-Drug Interactions. Molecules. 2016;21(8):1052. doi:10.3390/molecules21081052

21. Golej J, Boigner H, Burda G, Hermon M, Trittenwein G. Severe respiratory failure following charcoal application in a toddler. Resuscitation. 2001;49(3):315-318. doi:10.1016/s0300- 9572(00)00362-2

22. Picon PD, Picon RV, Costa AF, et al. Randomized clinical trial of a phytotherapic compound containing Pimpinella anisum, Foeniculum vulgare, Sambucus nigra, and Cassia augustifolia for chronic constipation. BMC Complementary and Alternative Medicine. 2010;10(1). doi:10.1186/1472-6882-10-17.

23. Dubey P, Jayasooriya AP, Cheema SK. Fish oil induced hyperlipidemia and oxidative stress in BioF1B hamsters is attenuated by elderberry extract. Applied Physiology, Nutrition, and Metabolism. 2012;37(3):472-479. doi:10.1139/h2012-030.

24. Vauzour D, Tejera N, Oneill C, et al. Anthocyanins do not influence long-chain n-3 fatty acid status: studies in cells, rodents and humans. The Journal of Nutritional Biochemistry. 2015;26(3):211-218. doi:10.1016/j.jnutbio.2014.09.005.

25. Levine WZ, Samuels N, Sheshet MEB, Grbic JT. A Novel Treatment of Gingival Recession using a Botanical Topical Gingival Patch and Mouthrinse. The Journal of Contemporary Dental Practice. 2013;14:948-953. doi:10.5005/jp-journals-10024-1431.

26. Harokopakis E, Albzreh MH, Haase EM, Scannapieco FA, Hajishengallis G. Inhibition of Proinflammatory Activities of Major Periodontal Pathogens by Aqueous Extracts From Elder Flower (Sambucus nigra). Journal of Periodontology. 2006;77(2):271-279. doi:10.1902/jop.2006.050232.

27. Zakay-Rones Z, Thom E, Wollan T, Wadstein J. Randomized Study of the Efficacy and Safety of Oral Elderberry Extract in the Treatment of Influenza A and B Virus Infections. Journal of International Medical Research. 2004;32(2):132-140. doi:10.1177/147323000403200205.

28. Zakay-Rones Z, Varsano N, Zlotnik M, et al. Inhibition of Several Strains of Influenza Virus in Vitro and Reduction of Symptoms by an Elderberry Extract (Sambucus nigra L.) during an Outbreak of Influenza B Panama. The Journal of

Alternative and Complementary Medicine. 1995;1(4):361-369. doi:10.1089/acm.1995.1.361.

29. Chrubasik C, Maier T, Dawid C, et al. An observational study and quantification of the actives in a supplement withSambucus nigraandAsparagus officinalisused for weight reduction. Phytotherapy Research. 2008;22(7):913-918. doi:10.1002/ptr.2415.

30. Lee Y-M, Yoon Y, Yoon H, Park H-M, Song S, Yeum K-J. Dietary Anthocyanins against Obesity and Inflammation. Nutrients. 2017;9(10):1089. doi:10.3390/nu9101089.

Elderberries and the Glycemic Load

By Zora DeGrandpre, ND

The elderberry—more formally known as Sambucus nigra spp—grows as a small, deciduous shrub native to North America, Europe, Asia and North Africa. It has been used as both a food and a medicine for thousands of years. Elderberries are a very low glycemic load food that is a healthy addition to any diet and is also appropriate for individuals with diabetes. According to some research, eating elderberries can form a useful plan for weight loss and potentially for the prevention of diabetes.[1]

Elderberry Nutrition: Basic Facts

Elderberries are a low calorie, low-fat food that is considered rich in fiber and complex carbohydrates. [2] Elderberries also contain all the essential amino acids—these are the protein building blocks that we need to obtain from foods because the human body cannot make (synthesize) these amino acids. [3]

Elderberries are a source of Vitamin C, providing 87% of the RDA. Elderberries also contain Vitamin A (18% of RDA), Vitamin B6 (17% RDA) as well as iron (13% RDA). Elderberries also contain a significant amount of other B vitamins and minerals including calcium, magnesium, phosphorus, selenium, copper, and zinc.[4]

Glycemic Index and Glycemic Load

The term "sugar" describes a class of chemical substances while the term "carbohydrate" includes sugars, starch, and fiber. Table sugar (sucrose) is one type.

While elderberries contain both simple and complex sugars, elderberries have a very low glycemic load—it is 1 for a 1 ounce (28g) serving. It may be surprising to some, especially those people with diabetes, but elderberries and other dark berries have a very low GL because most of these sugars are bound up as complex sugars and bound up in the fiber of the fruit.[5]

In an animal study, elderberry extracts were fed to rats also fed a high-fat diet. The authors found that use of the elderberry extract resulted in lower blood glucose and lower blood fat levels. [6] Another animal study concluded that elderberry extract reduced inflammation, increased insulin sensitivity and stabilized metabolism in obese mice.[7] To see how the elderberry compares with other dark berries, see Table 1. (To give a quick comparison, a baked potato has a GL of 33 and white rice has a GL of 35.)

Table 13: Comparison of the Glycemic Load and Carbohydrates Per Gram of Berries from Concentrated Apple Juice. Data from Ref. #2

DARK BERRY (AS RAW FOOD)	GLYCEMIC LOAD	CARBOHYDRATES (GM)/ GRAM OF BERRIES OR JUICE
Elderberries (28 g/1 ounce)	1	0.184 g
Blueberries (28 g/1 ounce)	1	0.145 g
Cranberries (28 g/1 ounce)	0	0.109 g
Blackberries (28 g/1 ounce)	1	0.098 g
Cherries, dark (28 g/1 ounce)	1	0.160 g
Concentrated Apple juice (28 g/1 ounce)*	6	0.115 g

* Concentrated apple juice has a higher GL because the sugars in apple juice are simple sugars and are absorbed by the body more quickly.

Since elderberries and other dark berries have such a low glycemic load, they can be eaten as a between-meal snack or as part of a regular meal. When elderberry juice is combined with apple syrup, we can estimate that the glycemic load will be somewhere between the glycemic load of elderberries and apple syrup. While it's not syrup, concentrated apple juice can be used as an approximation of apple syrup, so as a first estimate, the glycemic load of a mixture of elderberry juice and apple syrup would be between 1 and 6. In addition, the greater the percentage of elderberry juice, the GL would be closer to 1. It should be emphasized that both these values are very low— remember, any GL less than ten is considered very low.

Conclusions

Elderberries are a highly nutritious food, with a very low GL. Elderberry juice mixed with apple syrup can also be reasonably seen as having a very low GL as well. As such, elderberries and elderberry juice would be appropriate for diabetic individuals as well as those who are interested in following a low GL-based diet to help them lose weight. [8]

In addition, elderberries contain some of the highest levels of antioxidant flavonoids, plant substances that are being studied for both the prevention and treatment of several human diseases.

Photo Credit: Shutterstock

REFERENCES

1. Sidor, Andrzej, and Anna Gramza-Michałowska. "Advanced research on the antioxidant and health benefit of elderberry (Sambucus nigra) in food–a review." Journal of functional foods 18 (2015): 941-958.

2. Sidor, Andrzej, and Anna Gramza-Michałowska. "Advanced research on the antioxidant and health benefit of elderberry (Sambucus nigra) in food–a review." Journal of functional foods 18 (2015): 941-958.

3. https://ndb.nal.usda.gov/ndb/foods/show/2200

4. http://nutritiondata.self.com/facts/fruits-and-fruit-juices/1883/2

5. Sidor, Andrzej, and Anna Gramza-Michałowska. "Advanced research on the antioxidant and health benefit of elderberry (Sambucus nigra) in food–a review." Journal of functional foods 18 (2015): 941-958.

6. Salvador, Ângelo C., et al. "Effect of Elderberry (Sambucus nigra L.) Extract Supplementation in STZ-Induced Diabetic Rats Fed with a High-Fat Diet." International journal of molecular sciences 18.1 (2016): 13.

7. Farrell, Nicholas J., et al. "Black elderberry extract attenuates inflammation and metabolic dysfunction in diet-induced obese mice." British Journal of Nutrition 114.8 (2015): 1123-1131.

8. Cunha, Sara, Diana Meireles, and Jorge Machado. "Sambucus nigra–a promising natural source for human health." Exp Pathol Health Sci 8 (2016): 59-66

Are Elderberries Safe for Kids?

By Dr. Pierrette Mimi Poinsett, MD, Pediatrician

Medical science has neither evaluated the efficacy nor the safety of elderberry supplementation in children. To-date, there have been no clinical studies of elderberry usage in children.

I searched for data about elderberry use in childhood in the following databases: Google Scholar, Lactmed, PubMed, Trip Database and Europe PMC covering the period 1984 to present. The most recent articles mentioning children are from 2013 and 2014 (1, 2). The 2013 and 2014 references were a systematic review of research from 1970 to 2013. No clinical studies of children were noted in both review articles. There are no clinical studies in the scientific literature involving children from 2014 to present.

Cooked elderberries are made into jams, pies, and wine and are thus safe as food. Raw elderberry can cause nausea and vomiting. The leaves, branches, and flowers contain a chemical that is poisonous (1). Elderberry extract and juice are sold as dietary supplements either as a pure supplement or in combination with other ingredients.

There are limited studies of elderberry in adults (1,2). In these studies, safety has been established for non-pregnant non- breastfeeding adults. Elderberry also has not been studied in pregnant and breastfeeding women- another source of exposure to infants. There are no evidence-based recommendations for the use of elderberry in pregnant and breastfeeding women (3,4).

REFERENCES

1. European Medicines Agency, Assessment report on Sambucus nigra L., fructus, 2013 retrieved from http://www.ema.europa.eu/docs/en_GB/document_library/ Herbal_-_HMPC_assessment_report/2013/04/WC500142245.pdf

2. Ulbricht, C, Basch, E, Cheung L et al., An Evidence-Based Systematic Review of Elderberry and Elderflower (Sambucus nigra)by the Natural Standard Research Collaboration, J of Dietary Suppl, 2014 retrieved from https://www. researchgate.net/publication/259696401_An_Evidence- Based_Systematic_ Review_of_Elderberry_and_Elderflower_Sambucus_nigra_by_the_Natural_Sta ndard_Research_Collaboration

3. Holst, L, Havnen, G.C. and Nordeng, H., Echinacea and elderberry-should they be used against upper respiratory tract infections during pregnancy? 2014, Front Pharmacol 5: 31. Retrieved from https://www.ncbi.nlm.nih.gov/pmc/articles/ PMC3941201/

4. Lactmed, Elderberry. Retrieved from https://toxnet.nlm.nih.gov/cgi-bin/sis/ search2

Quick Read — Article Summaries

By John Stack, DVM

The Key to Elderberry's Health Benefits

The benefits of a diet high in fruits and vegetables has been firmly established, with many of the health effects attributed to high antioxidant content. Quantifying the antioxidant content of foods, however, can be quite challenging and is often non-standardized. Laboratory-based approaches to measuring antioxidant content do not correlate well to how they are processed or absorbed in the body.

Additionally, there are multiple approaches to measuring antioxidants and there is disagreement among scientists about the validity of these various approaches.

Despite this, research does support the notion that a diet rich in fruits and vegetables will result in increased antioxidant activity in the body, and polyphenols in particular are anti-inflammatory and protective of the cardiovascular system. Elderberries are rich in multiple polyphenols and are a good source for people trying to add more antioxidants in their diets.

A Closer Look at Elderberry's Polyphenols

Polyphenols are the colorful antioxidant compounds found in many fruits and in berries. Elderberries contain a specific category of polyphenols called anthocyanins, which give them their deep blue-purple coloration.

Polyphenols have undergone extensive study for their potential protective effects against cancer, heart disease, and inflammatory conditions like osteoarthritis.

Although there is not definitive evidence to suggest that the anthocyanins found in elderberries have more benefits than polyphenols found in other produce, the fact that elderberries contain such a high amount of these beneficial compounds make them an excellent dietary source. Although processed produce still contains polyphenols, raw or minimally-processed products will have the highest sources.

The Elderberry - "The Tree of Music" - is a Rich Source of Anti-oxidants

Elderberry fruit has a long history of use as a medicinal plant by Native Americans. We now know that elderberries have a high anti-oxidant content, mostly in the form of polyphenols (in particular anthocyanidins). These protect the body against free radicals – charged molecules which damage DNA and cell membranes in the body.

Anti-oxidants found in elderberries and other berries scavenge these free radicals and render them harmless, sparing the body from damage.

Elderberries are also potent inhibitors of an enzyme called cyclo-oxygenase-2 (COX- 2), which plays a key role in inflammation within the body. Many prescription pharmaceutics target COX-2 to relieve pain and swelling associated inflammatory conditions such as arthritis.

The other anti-oxidants found in elderberries may also have protective effects against cancer, cardiovascular disease, and degenerative diseases – properties which were perhaps observed by early Native Americans.

Does the Nutritional Content of Berries Vary Depending on Whether They Were Grown Organically or Conventionally?

There is little information about the nutritional content of organically vs conventionally-grown elderberries, however several studies on blueberries, strawberries, and raspberries have shown that the organically-grown berries have higher anti-oxidant content compared to the conventionally-grown counterparts.

The reasons behind this are not completely clear, but it has been hypothesized that organically-grown plants generate higher levels of anti-oxidants as a defensive/ protective mechanism, while conventionally-grown plants are externally protected by fungicides, and pesticides, resulting in a decreased need for these compounds.

That said, the nutritional difference between organic and conventional berries is generally negligible. The chemicals used in conventional practices, however, may be of greater concern. Washing the berries can help reduce the amount of pesticides and fungicides, however, even after thorough, 5-minute washes and use of detergents, detectable levels of these chemicals may remain. Although organic and wild-grown berries may not have these chemicals, it is important to consider whether or not they were grown near areas of pollution, such as near roads, as this may also result in chemical contamination.

How Good a Source of Polyphenols is Elderflower Tea?

Although the fruit of the elderberry plant receives most of the attention from antioxidant seekers, elderflower also has a high polyphenol content, though they are different than those found in the berries. Furthermore, it may take less of elderflower polyphenols to achieve the same anti-inflammatory effects as the polyphenols found in elderberries.

Tea made from elderflower has a very high concentration of these polyphenols, and compares well against other herbal teas such as chamomile. Overall, it appears both the berry and the flower from the elder tree provide powerful, but different, polyphenols.

Maximizing the Antioxidant Power of Berries

Antioxidant levels vary widely across different types of berries; however, elderberries are among the berries with the highest antioxidant, mostly in the form of anthocyanins. Antioxidant levels are impacted by where and how the berries were grown, but are also impacted by storage and processing. For a short time, anthocyanin levels may actually increase at room temperature or when frozen, however they will begin to decrease after this.

Exposure to high heat (as often occurs in making jams) can reduce antioxidant content, but it may simultaneously make the antioxidants easier for the body to absorb, potentially off-setting the overall loss in content. Lower levels of heat that occur normally in baking and pan-frying do not seem to impact antioxidant levels, though it may change the profile of antioxidants (for example, degrading anthocyanins, but releasing other antioxidant compounds).

Freezing and drying also does not significantly impact anti-oxidants. The best approach to maximize the benefits of antioxidants in berries is to enjoy them soon after purchase, regardless of culinary intent.

Are Elderberries Safe to Eat? Raw Heated, or Otherwise Processed?

In general, ripe fruit and flowers from the elderberry plant are considered safe for consumption, however other parts of the plant such as the bark, leaves, seeds, and unripe fruit may contain cyanide-based compounds, especially if eaten raw. In regards to the fruit and flowers, there are some considerations one should make before consuming anything. Plants that grow in polluted areas (such as near roads or industrial run-off sites) may have higher levels of heavy metals or other toxic compounds.

Additionally, it is possible that plant material may be contaminated with bacteria or other disease-causing organisms. Washing, heating, and other processing of elderberries can reduce the risk of exposure to these organisms, however it also comes at the cost of losing some of the beneficial compounds found in elderberries. Cooking can also reduce the risk of cyanide exposure in the remaining parts of the plant, making them safe for limited consumption, however it may not completely eliminate all of the toxic compounds.

Elderberry Skins: Are They Important?

Many fruits have nutrients in the pulp as well as the skin (or peel), and sometimes the nutritional profiles of each can vary widely, and on a "by weight" basis, skin may sometimes have a higher density of nutrients. This point is often moot, however, given that there usually is a much higher ratio of pulp to skin in most fruits. For elderberries, most studies have evaluated whole extracts or juices, and have not directly assessed the skin vs the pulp.

It likely makes little difference whether elderberries are consumed as whole berries or as a juice, as either will be very high in antioxidants.

Why the Whole Berry Works Better Than a Pill

The temptation to take polyphenol supplements – which offer convenience and high concentrations of polyphenols – can be very strong. In some cases, supplementation with herbal extracts can be helpful in particular scenarios, however for the general population, the benefits of supplements are often unclear, or in some cases, potentially harmful.

The strongest evidence for benefits from polyphenols is usually associated with a diet rich in fruits and vegetables, not from individual supplementation. There are likely complex interactions between the many antioxidant compounds found in plant products and the human body. Obtaining a diverse range of these compounds by eating whole fruits and vegetables is preferable and more sensible than relying on a specific supplement or extract alone.

Are Elderberry Polyphenols Destroyed by Heat?

Cooking and processing can have variable effects on polyphenols. In some instances, polyphenols degrade, while in others, they become more available for the body. Sometimes, the processing method may even increase polyphenol content, as is the case with blueberries baked in the presence of yeast.

Studies on cooked Dwarf elderberry found a dramatic reduction in one polyphenol in particular (cyanidin-3-glucoside), however, total polyphenol content only changed 5%, as the majority of other compounds are more stable. Thus, although maximal polyphenol levels will probably be found in unprocessed berries, there is still a high content in prepared or heated berries as well.

An Overview of Elderberry's Health Benefits

Elderberries have been regarded as having medicinal qualities since Hippocrates, the ancient Greek who is considered the "father of medicine." We are only beginning to understand why elderberries have held such high esteem throughout history, but so far, research has found numerous health benefits. The most well-researched effect of elderberries is against viral infections; specifically, the influenza virus. Besides directly blocking the virus from entering mammalian cells, elderberry also appears to activate the immune system against infections.

Elderberries are also well known for their antioxidant properties, which may be protective against the effects of ageing as well as preventing the development of some cancers. More and more, research appears to support the ancient notions of the medicinal qualities of elderberries.

A Comprehensive Review of Elderberry's Effects

There is a broad array of medicinal effects which are attributed to elderberries. Many of these purported effects have undergone scientific scrutiny. The areas which have the most scientific support tend to be in the context of shortening the duration and severity of flu and other respiratory infections. There is also evidence that elderberry is generally anti-inflammatory, and shows promise in the treatment of conditions with inflammation such as gingivitis.

There are also positive metabolic effects of elderberries, including decreased circulating triglycerides, improvements in cholesterol levels, and decreased atherosclerosis. Elderberry also appears to be an effective laxative to relieve constipation. Other claims such as diuretic effects and improving weight loss have so far not been scientifically demonstrated. Elderberries are being examined for use in a range of other situations, including cancer and wound-healing. Regardless of their medicinal effects, they have well-known anti-oxidant properties, and may contribute to a healthy diet even in the absence of specific medicinal qualities.

Can Elderberries Cure the Flu?

Elderberries as a remedy against influenza is a promising, if incomplete, area of research. Multiple double-blind, placebo-controlled studies have consistently shown that use of elderberry products at the onset of flu infections can shorten the duration or severity of illness in some people. Although these raise hopes for elderberry use in influenza, there are several caveats. The first caveat is that all of the studies were relatively small, making them less powerful than large, clinically-controlled trials.

Second, different products were used across studies, and some of these products were mixed herbal products, which contained ingredients other than elderberries, or even products with undisclosed formulations. Although promising and deserving of further research, the available evidence can only be considered, at best, preliminary.

Which Elderberry Products are Best for the Flu?

When it comes to fighting the flu, there are a wide variety of elderberry options to choose from. The most well-researched commercial product is Sambucol, which contains a 38% elderberry extract. Other commercial products tend to contain less elderberry extract and a higher proportion of other ingredients. It is not known whether a higher proportion of elderberry extract is superior to products which contain lower concentrations, as the other ingredients may also play a role.

Juices, jams, and jellies are likely to contain more active compounds because they have undergone less processing, however capsules and tablets may still provide some benefit as well.

Elderberry v. Tamiflu: Which is Right for You?

Natural products such as elderberry are popular choices for fighting the flu, however there is an antiviral medication called Tamiflu (oseltamivir) which has been clinically shown to decrease the duration and severity of illness from the flu. There is only one study comparing Tamiflu against an elderberry product, however this single study showed the patients treated with elderberry/echinacea extract recovered from their infections more quickly and with fewer side effects.

Tamiflu has a very specific mechanism of action (neuraminidase inhibition) while elderberry is thought to have several different mechanisms, including blocking infection, improving the immune system, and working as an antioxidant. Tamiflu is significantly more expensive than most elderberry preparations, and comes with side effects such as nausea, vomiting, and headaches. More research is needed for a better comparison; however, elderberry seems to stack up favorably against Tamiflu.

Overview of EU Regulation of Elderberry as Medicine

Elderberry is regulated as a medicinal product in the European Union by the European Medicines Agency (EMA), a regulatory body similar to the FDA in the United States. The EMA allows elderberry to be sold and consumed as natural product under a "traditional use" registration, which means there is a lower standard of evidence about its use as a medicine, however it has a long history of use with plausible efficacy.

It can be sold and used without a doctor's supervision, however it is generally not offered as a treatment by doctors. There is a category of regulation with a higher burden of evidence, called "well-established use," which would allow it to be offered by physicians for specific conditions. The EMA publishes assessment reports of every drug marketed within the EU, including elderberry, which is updated every 5 years, and is made available to the general public.

These reports contain information about the safety, efficacy, applications, product, and quality. Currently, it is unclear if elderberry's status as a "traditional use" medication will change in the future.

How Much Elderberry Should I Take? What Should the Dosage be?

Proper dosing is important for any product taken for medicinal or supplemental purposes. At this point, there has been no "proven" dose of elderberry when taken for medicinal purposes. There have been a wide range of does described in clinical trials, however, depending on the product and purpose of the supplementation.

When using prepared products, it is probably best to stick to the labeled dose, as this is what was likely tested in trials. For juices and extracts, it is even less clear due to variability in concentrations, and there is not currently reliable evidence to suggest a dosing scheme.

Current Scientific Understanding of the Side Effects and Interactions from Consuming Elderberry

The potential side effects of any medication or supplement should always be considered. Although generally safe, elderberry plants do contain cyanide-containing compounds which can make a person ill if consumed. They also contain ribosome-inactivating proteins (RIPs) which can be toxic to mammalian cells, however this is likely not a major issue for humans, unless a person is allergic to them. There may be concerns when elderberry is combined with other medications, however.

Elderberry's impact on blood sugar could become an issue for a diabetic who takes insulin, potentially predisposing to hypoglycemia. Elderberry also may interfere with pain medications and anesthetic agents. Immunocompromised individuals should also exercise caution, due to the uncertain manner in which elderberry interacts with the immune system.

Although elderberry generally appears safe, carries few side effects, and is potentially beneficial as a supplement, there is a paucity of information regarding its interactions with many medications. It should be used judiciously and dosing guidelines labeled on prepared products should be followed.

Elderberries and the Glycemic Load

Management of blood sugar levels can be an important factor in weight regulation and risk of diabetes. Some research suggests that elderberries may be a good option for people who are concerned about "glycemic load," a measure of how quickly sugar is absorbed into the blood stream after eating.

Elderberries are rich in complex sugars and fiber, which leads to slow absorption and a more controlled rise in blood sugar levels. Animal studies have found elderberry extract improves insulin sensitivity and decreased circulating blood triglycerides as well. These qualities make them a nutritious option for people with diabetes, or those who struggle with regulating their body weight.

Are Elderberries Safe for Kids?

When properly prepared, elderberries are considered safe as food items (such as pies, jams, jellies, etc.) for anyone. The question of supplementation, however, is a different matter. Supplements are often used for medicinal purposes, and may contain a mixture of ingredients in addition to elderberry.

Although elderberry supplements have been found to be safe in adults, there are no studies which have evaluated the use of elderberry supplements and/or remedies in children, or in pregnant or breastfeeding women. Before beginning any supplementation in children, it is best to discuss the reasons for supplementation and specific supplements with a pediatrician.

Elderberries Benefit People - But What About Pets?

What's Inside?

Over 14 pages of clearly written, well-researched conclusions about elderberries's documented impacts on animal health.

Topics covered include:

- *Has Elderberry Been Studied In Animals?*
- *Does is Work? In What Ways?*
- *What Dose is Acceptable?*
- *Are side-effects possible?*
- And more!

Disclosures & Disclaimers

This book is published in both paperback and electronic format. It was written by various contributors for the Publisher as a work-for-hire. The Publisher is the sole copyright owner.

The Publisher does not make any claim to the intellectual property rights of third-party vendors, their subsidiaries, or related entities. All trademarks and service marks are the properties of their respective owners. All references to these properties are made solely for editorial purposes. Except for marks actually owned by the Publisher, no commercial claims are made to their use, and the Publisher is not affiliated with such marks in any way.

Unless otherwise expressly noted, none of the individuals or business entities mentioned herein has endorsed the contents of this book.

Health Disclaimers

As an express condition to reading this book, you understand and agree to the following terms.

This book is a general educational health-related information product. This book does not contain medical advice.

The book's content is not a substitute for direct, personal, professional medical care and diagnosis. None of the exercises or treatments (including products and services) mentioned in this book should be performed or otherwise used without prior approval from your physician or other qualified professional health care provider.

There may be risks associated with participating in activities or using products and services mentioned in this book for people in poor health or with pre-existing physical or mental health conditions.

Because these risks exist, you will not use such products or participate in such activities if you are in poor health or have a pre-existing mental or physical condition. If you choose to participate in these risks, you do so of your own free will and accord, knowingly and voluntarily assuming all risks associated with such activities.

Limits of Liability & Disclaimers of Warranties

Because this book is a general educational information product, it is not a substitute for professional advice on the topics discussed in it.

The materials in this book are provided "as is" and without warranties of any kind either express or implied. The Publisher disclaims all warranties, express or implied, including, but not limited to, implied warranties of merchantability and fitness for a particular purpose. The Publisher does not warrant that defects will be corrected, or that any website or any server that makes this book available is free of viruses or other harmful components. The Publisher does not warrant or make any representations regarding the use or the results of the use of the materials in this book in terms of their correctness, accuracy, reliability, or otherwise.

Applicable law may not allow the exclusion of implied warranties, so the above exclusion may not apply to you.

Under no circumstances, including, but not limited to, negligence, shall the Publisher be liable for any special or consequential damages that result from the use of, or the inability to use this book, even if the Publisher, or an authorized representative has been advised of the possibility of such damages. Applicable law may not allow the limitation or exclusion of liability or incidental or consequential damages, so the above limitation or exclusion may not apply to you. In no event shall the Publisher's total liability to you for all damages, losses, and causes of action (whether in contract, tort, including but not limited to, negligence or otherwise) exceed the amount paid by you, if any, for this book.

You agree to hold the Publisher of this book, principals, agents, affiliates, and employees harmless from any and all liability for all claims for damages due to injuries, including attorney fees and costs, incurred by you or caused to third parties by you, arising out of the products, services, and activities discussed in this book, excepting only claims for gross negligence or intentional tort.

You agree that any and all claims for gross negligence or intentional tort shall be settled solely by confidential binding arbitration per the American Arbitration Association's commercial arbitration rules and judgment on the award rendered by the arbitrator(s) may be entered in any court having jurisdiction thereof. All arbitration must occur in New London County, Connecticut USA. You agree that you will not aggregate your claim with third party claims. Arbitration fees and costs shall be split equally, and you are solely responsible for your own lawyer fees.

Facts and information are believed to be accurate at the time they were placed in this book. All data provided in this book is to be used for information purposes only. The information contained within is not intended to provide specific legal, financial, tax, physical or mental health advice, or any other advice whatsoever, for any individual or company and should not be relied upon in that regard. The services described are only offered in jurisdictions where they may be legally offered. Information provided is not all-inclusive, and is limited to information that is made available and such information should not be relied upon as all-inclusive or accurate.

For more information,please contact the Publisher by e-mail at bill@elderberryguru. com or by mail at P.O. Box 353, Old Lyme, CT 06371 USA.

IF YOU DO NOT AGREE WITH THESE TERMS AND EXPRESS CONDITIONS, DO NOT READ THIS BOOK. YOUR USE OF THIS BOOK, PRODUCTS, SERVICES, AND ANY PARTICIPATION IN ACTIVITIES MENTIONED IN THIS BOOK, MEAN THAT YOU ARE AGREEING TO BE LEGALLY BOUND BY THESE TERMS.

Affiliate Compensation & Material Connections Disclosure

This book may contain hyperlinks to websites and information created and maintained by other individuals and organizations. The Publisher does not control or guarantee the accuracy, completeness, relevance, or timeliness of any information or privacy policies posted on these linked websites.

You should assume that all references to products and services in this book are made because material connections exist between the Publisher and the providers of the mentioned products and services (each a "Provider"). You should also assume that all hyperlinks within this book are affiliate links for the Publisher.

The Publisher recommends products and services in this book based in part on a good faith belief that the purchase of such products or services will help readers in general.

The Publisher has this good faith belief because (a) the Publisher has tried the product or service mentioned prior to recommending it or (b) the Publisher has researched the reputation of the Provider and has made the decision to recommend the Provider's products or services based on the Provider's history of providing these or other products or services.

The representations made by the Publisher about products and services reflect the Publisher's honest opinion based upon the facts known to the Publisher at the time this book was published.

Because there may be a material connection between the Publisher and Providers of products or services mentioned in this book, you should always assume that the Publisher may be biased because of the Publisher's relationship with a Provider and/or because the Publisher has received or will receive something of value from a Provider.

Made in the USA
Middletown, DE
04 September 2019